Scholastic Success With Tests:
Math Workbook

Grade 5

by Michael Priestley

New York • Toronto • London • Auckland • Sydney •
Mexico City • New Delhi • Hong Kong • Buenos Aires

Cover design by Maria Lilja
Cover art by Victoria Raymond
Interior design by Creative Pages Inc.
Interior illustrations by Kate Flanagan

ISBN 0-439-42569-7

Contents

Introduction

In this book, you will find eight Practice Tests designed to help students prepare to take standardized tests. Each test has 20–30 multiple-choice items that closely resemble the kinds of questions students will have to answer on "real" tests. Each part of the test will take 30–40 minutes for students to complete.

The Math skills measured in these tests and the types of questions are based on detailed analyses and correlations of the five most widely used standardized tests and the curriculum standards measured by many statewide tests, including the following:

Stanford Achievement Test
CTBS TerraNova
Metropolitan Achievement Test
Iowa Test of Basic Skills
California Achievement Test

California's STAR Test
TAAS (Texas)
MCAS (Massachusetts)
FCAT (Florida)
New York

How to Use the Tests

Tell students how much time they will have to complete the test. Encourage students to work quickly and carefully and to keep track of the remaining time—just as they would in a real testing session. You may have students mark their answers directly on the test pages, or you may have them use a copy of the **Answer Sheet**. An answer sheet appears at the end of each test. The answer sheet will help students become accustomed to filling in bubbles on a real test. It may also make the tests easier for you to score.

We do not recommend the use of calculators. For Practice Tests 2 and 6, students will need an inch ruler and a centimeter ruler to answer some of the questions.

At the back of this book, you will find **Tested Skills** charts and **Answer Keys** for the eight Practice Tests. The Tested Skills charts list the skills measured in each test and the test questions that measure each skill. These charts may be helpful to you in determining what kinds of questions students answered incorrectly, what skills they may be having trouble with, and who may need further instruction in particular skills. To score a Practice Test, refer to the Answer Key for that test. The Answer Key lists the correct response to each question.

To score a Practice Test, go through the test and mark each question answered correctly. Add the total number of questions answered correctly to find the student's test score. To find a percentage score, divide the number answered correctly by the total number of questions. For example, the percentage score for a student who answers 20 out of 25 questions correctly is $20 \div 25 = 0.80$, or 80%. You might want to have students correct their own tests. This will give them a chance to see where they made mistakes and what they need to do to improve their scores on the next test.

On the next page of this book, you will find **Test-Taking Tips**. You may want to share these tips and strategies with students before they begin working on the Practice Tests.

Test-Taking Tips: Mathematics

1. For each part of the test, read the directions carefully so you know what to do. Then read the directions again—just to make sure.

2. Look for key words and phrases to help you decide what each question is asking and what kind of computation you need to do. Examples of key words: *less than, greatest, least, farther, longest, divided equally.*

3. To help solve a problem, write a number sentence or equation.

4. Use scrap paper (or extra space on the test page) to write down the numbers and information you need to solve a problem.

5. If a question has a picture or diagram, study it carefully. Draw your own picture or diagram if it will help you solve a problem.

6. Try to solve each problem before you look at the answer choices. (In some tests, the correct answer may be "Not Given" or "Not Here," so you will want to be sure of your answer. In these Practice Tests, some of the Math questions use "NG" for "Not Given.")

7. Check your work carefully before you finish. (In many questions, you can check your answer by working backwards to see if the numbers work out correctly.)

8. If you are not sure which answer is correct, cross out every answer that you know is wrong. Then make your best guess.

9. To complete a number sentence or equation, try all the answer choices until you find the one that works.

10. When working with fractions, always reduce (or rename) the fractions to their lowest parts. When working with decimals, keep the decimal points lined up correctly.

Practice Test 1

Numeration and Number Concepts

Practice Test 1

Directions. Choose the best answer to each question. Mark your answer.

1. In 1999, people in the United States held about $1,755,000 in $5,000 bills. How is 1,755,000 written in words?
   - Ⓐ seventeen million five hundred fifty thousand
   - Ⓑ one hundred thousand seven hundred fifty-five
   - Ⓒ one million seven hundred fifty-five thousand
   - Ⓓ one thousand seven hundred fifty-five

2. The country of South Africa covers an area of four hundred seventy-one thousand nine miles. How is the area written as a numeral?
   - Ⓕ 471,009
   - Ⓖ 400,071,009
   - Ⓗ 400,710,009
   - Ⓙ 470,109

3. Which is an even number?
   - Ⓐ 2345
   - Ⓑ 5636
   - Ⓒ 3789
   - Ⓓ 1847

4. The chart shows the area of the four largest U.S. national parks, all of them in Alaska.

National Park	Area (acres)
Denali	4,740,912
Gates of the Arctic	7,523,898
Katmai	3,674,530
Wrangell-Saint Elias	8,323,618

Which national park is largest?
   - Ⓕ Denali
   - Ⓖ Gates of the Arctic
   - Ⓗ Katmai
   - Ⓙ Wrangell-Saint Elias

5. The chart lists the four most commonly spoken languages in U.S. homes, other than English.

Language	# of Speakers
French	1,702,000
German	1,547,000
Italian	1,309,000
Spanish	17,339,000

Which list shows the languages from greatest to least number of speakers?
   - Ⓐ French, German, Italian, Spanish
   - Ⓑ Spanish, French, German, Italian
   - Ⓒ Italian, German, French, Spanish
   - Ⓓ Spanish, Italian, German, French

GO ON

Practice Test 1 *(continued)*

6. 40,000 + 500 + 6 =
- Ⓕ 40,506
- Ⓖ 45,600
- Ⓗ 40,000,506
- Ⓙ 40,500,006

7. An organization has a sign at its national headquarters showing the number of members. One of the digits has fallen from the sign.

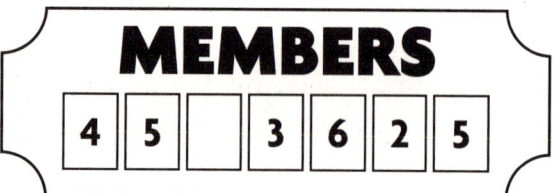

The missing digit represents what place value?
- Ⓐ hundreds
- Ⓑ thousands
- Ⓒ ten thousands
- Ⓓ hundred thousands

8. Lake Superior covers an area of about 31,820 square miles. What is that number rounded to the nearest ten thousand miles?
- Ⓕ 40,000
- Ⓖ 32,000
- Ⓗ 31,000
- Ⓙ 30,000

9. The Johnsons drove 1259 miles on their vacation in the Rocky Mountains. What is that number rounded to the nearest hundred miles?
- Ⓐ 1000
- Ⓒ 1300
- Ⓑ 1200
- Ⓓ 2000

10. 56, 47, 38, _____, 20, 11, 2 . . .

What number goes in the blank space in this number pattern?
- Ⓕ 30
- Ⓗ 27
- Ⓖ 29
- Ⓙ 20

11. Marta made this pattern of tiles on the kitchen floor.

Which section of tile will complete the pattern?

Ⓐ   Ⓒ

Ⓑ

GO ON ⇒

Practice Test 1 (continued)

12. Which of these must be an odd number?

- Ⓕ the sum of an odd number and an odd number
- Ⓖ the sum of an even number and an even number
- Ⓗ the sum of an odd number and an even number
- Ⓙ the product of an even number and an even number

13. At their annual bird count, the members of the Birders Association saw 3504 Canada geese, 1461 Brant geese, and 542 snow geese. *About* how many geese did they see in all?

- Ⓐ 5500
- Ⓑ 5000
- Ⓒ 4500
- Ⓓ 4000

14. Brandon has organized his music collection of 123 CDs. His sister has a collection of 56 CDs. *About* how much larger is Brandon's collection than his sister's?

- Ⓕ twice as large
- Ⓖ three times as large
- Ⓗ four times as large
- Ⓙ five times as large

15. What is the greatest common factor of 8, 12, and 24?

- Ⓐ 2
- Ⓑ 4
- Ⓒ 6
- Ⓓ 8

16. What is the least common multiple of 9 and 27?

- Ⓕ 3
- Ⓖ 27
- Ⓗ 54
- Ⓙ 243

17. Look at the number line.

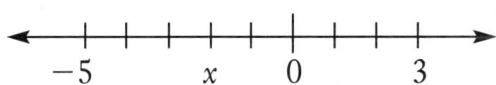

What number belongs in the place marked with an *x?*

- Ⓐ −4
- Ⓑ −3
- Ⓒ −2
- Ⓓ 2

GO ON

Practice Test 1 *(continued)*

18. Which number line shows the sum of −2 and 4?

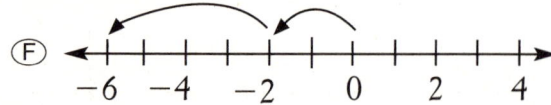

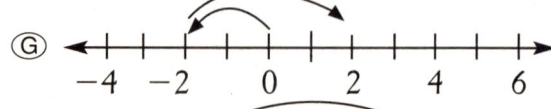

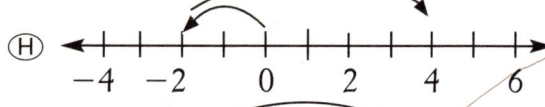

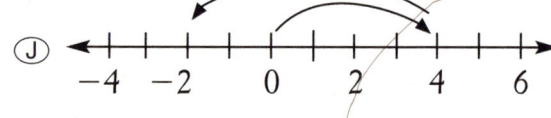

19. Alex has 8 gerbils, and 5 of them are light brown. What fraction of Alex's gerbils are light brown?

Ⓐ $\frac{3}{8}$

Ⓑ $\frac{3}{5}$

Ⓒ $\frac{5}{8}$

Ⓓ $\frac{3}{4}$

20. On Saturday, 45 of the 50 frogs' eggs in an aquarium hatched. What fraction of the frogs' eggs hatched?

Ⓕ $\frac{1}{10}$

Ⓖ $\frac{1}{5}$

Ⓗ $\frac{4}{5}$

Ⓙ $\frac{9}{10}$

21. Which trail through the Forbidden Swamp is longest?

Slimy Slope $3\frac{1}{3}$ miles
Hidden Hollow $3\frac{1}{2}$ miles
Mosquito's Misery $3\frac{1}{10}$ miles
Mucky Way $3\frac{3}{5}$ miles

Ⓐ Slimy Slope

Ⓑ Hidden Hollow

Ⓒ Mosquito's Misery

Ⓓ Mucky Way

22. In which list are the fractions arranged from least to greatest?

Ⓕ $\frac{9}{10}, \frac{4}{5}, \frac{6}{8}, \frac{6}{9}$ Ⓗ $\frac{4}{5}, \frac{6}{9}, \frac{6}{8}, \frac{9}{10}$

Ⓖ $\frac{6}{9}, \frac{6}{8}, \frac{4}{5}, \frac{9}{10}$ Ⓙ $\frac{4}{5}, \frac{6}{8}, \frac{6}{9}, \frac{9}{10}$

23. In a movie theater, 75 of the 225 seats are filled. What fraction of the seats are in use?

Ⓐ $\frac{1}{5}$ Ⓒ $\frac{2}{5}$

Ⓑ $\frac{1}{3}$ Ⓓ $\frac{2}{3}$

24. Which decimal number is equal to $\frac{2}{5}$?

Ⓕ 0.10   Ⓗ 0.30

Ⓖ 0.20   Ⓙ 0.40

GO ON

Practice Test 1 *(continued)*

25. At a track-and-field meet, the top 4 runners finished the 100-meter hurdles in the following times:

  18.84 seconds
  18.63 seconds
  18.70 seconds
  18.81 seconds

Which list shows the times in order from fastest to slowest?

Ⓐ 18.63, 18.70, 18.81, 18.84
Ⓑ 18.63, 18.70, 18.84, 18.81
Ⓒ 18.70, 18.81, 18.63, 18.84
Ⓓ 18.84, 18.81, 18.70, 18.63

26. Janice found the following prices for a 96-ounce jug of apple juice. How much does the least expensive jug of juice cost?

Ⓕ $3.69   Ⓗ $3.75
Ⓖ $3.96   Ⓙ $3.84

27. What number is shown on the number line?

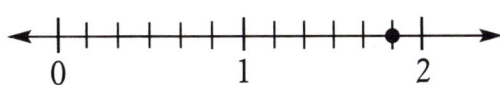

Ⓐ $\frac{5}{6}$

Ⓑ $1\frac{4}{5}$

Ⓒ $1\frac{5}{6}$

Ⓓ $2\frac{1}{6}$

28. This number line illustrates which number sentence?

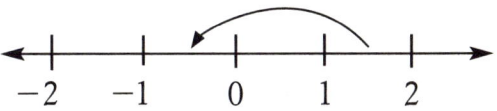

Ⓕ $2 - 0.5 = 1.5$
Ⓖ $-0.5 + 2 = 1.5$
Ⓗ $1.5 - 2 = -0.5$
Ⓙ $2 - 1.5 = 0.5$

29. Which expression is equivalent to $4 \times (y + 9)$?

Ⓐ $(4 \times y) + 9$
Ⓑ $4 \times y \times 9$
Ⓒ $4 + y + 9$
Ⓓ $(4 \times y) + (4 \times 9)$

30. $0 \times \frac{1}{2}(\frac{1}{4} + \frac{1}{2}) =$

Ⓕ 0

Ⓖ $\frac{1}{8} + \frac{1}{2}$

Ⓗ $0 + \frac{1}{4}$

Ⓙ $\frac{1}{8} + \frac{1}{4}$

STOP

ANSWER SHEET

Student Name _____ Grade _____

Teacher Name _____ Date _____

MATHEMATICS

1 Ⓐ Ⓑ Ⓒ Ⓓ Ⓔ	11 Ⓐ Ⓑ Ⓒ Ⓓ Ⓔ	21 Ⓐ Ⓑ Ⓒ Ⓓ Ⓔ	31 Ⓐ Ⓑ Ⓒ Ⓓ Ⓔ
2 Ⓕ Ⓖ Ⓗ Ⓙ Ⓚ	12 Ⓕ Ⓖ Ⓗ Ⓙ Ⓚ	22 Ⓕ Ⓖ Ⓗ Ⓙ Ⓚ	32 Ⓕ Ⓖ Ⓗ Ⓙ Ⓚ
3 Ⓐ Ⓑ Ⓒ Ⓓ Ⓔ	13 Ⓐ Ⓑ Ⓒ Ⓓ Ⓔ	23 Ⓐ Ⓑ Ⓒ Ⓓ Ⓔ	33 Ⓐ Ⓑ Ⓒ Ⓓ Ⓔ
4 Ⓕ Ⓖ Ⓗ Ⓙ Ⓚ	14 Ⓕ Ⓖ Ⓗ Ⓙ Ⓚ	24 Ⓕ Ⓖ Ⓗ Ⓙ Ⓚ	34 Ⓕ Ⓖ Ⓗ Ⓙ Ⓚ
5 Ⓐ Ⓑ Ⓒ Ⓓ Ⓔ	15 Ⓐ Ⓑ Ⓒ Ⓓ Ⓔ	25 Ⓐ Ⓑ Ⓒ Ⓓ Ⓔ	35 Ⓐ Ⓑ Ⓒ Ⓓ Ⓔ
6 Ⓕ Ⓖ Ⓗ Ⓙ Ⓚ	16 Ⓕ Ⓖ Ⓗ Ⓙ Ⓚ	26 Ⓕ Ⓖ Ⓗ Ⓙ Ⓚ	36 Ⓕ Ⓖ Ⓗ Ⓙ Ⓚ
7 Ⓐ Ⓑ Ⓒ Ⓓ Ⓔ	17 Ⓐ Ⓑ Ⓒ Ⓓ Ⓔ	27 Ⓐ Ⓑ Ⓒ Ⓓ Ⓔ	37 Ⓐ Ⓑ Ⓒ Ⓓ Ⓔ
8 Ⓕ Ⓖ Ⓗ Ⓙ Ⓚ	18 Ⓕ Ⓖ Ⓗ Ⓙ Ⓚ	28 Ⓕ Ⓖ Ⓗ Ⓙ Ⓚ	38 Ⓕ Ⓖ Ⓗ Ⓙ Ⓚ
9 Ⓐ Ⓑ Ⓒ Ⓓ Ⓔ	19 Ⓐ Ⓑ Ⓒ Ⓓ Ⓔ	29 Ⓐ Ⓑ Ⓒ Ⓓ Ⓔ	39 Ⓐ Ⓑ Ⓒ Ⓓ Ⓔ
10 Ⓕ Ⓖ Ⓗ Ⓙ Ⓚ	20 Ⓕ Ⓖ Ⓗ Ⓙ Ⓚ	30 Ⓕ Ⓖ Ⓗ Ⓙ Ⓚ	40 Ⓕ Ⓖ Ⓗ Ⓙ Ⓚ

Practice Test 2

Geometry and Measurement

Practice Test 2

Directions. Choose the best answer to each question. Mark your answer.

1. Which figure is a hexagon?

 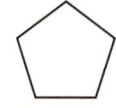

2. Mr. Blake's driveway is 9 yards long. How many feet is that?
- Ⓕ 3 ft
- Ⓖ 12 ft
- Ⓗ 18 ft
- Ⓙ 27 ft

3. Which figure shows a line of symmetry?

 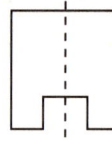   

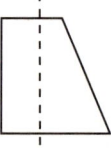   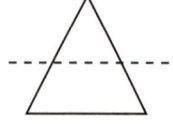

4. Melanie made a diagram of her rectangular vegetable garden. What is the area of her garden?

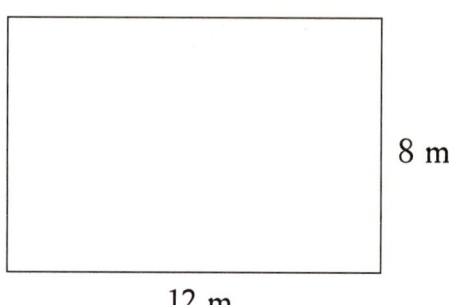

12 m

8 m

- Ⓕ 20 m²
- Ⓖ 40 m²
- Ⓗ 96 m²
- Ⓙ 192 m²

5. This table shows the costs for overnight delivery of packages.

Overnight Delivery	
Package Weight	**Rate**
0 – 1.0 lb	$13.95
1.1 – 2.0 lb	$16.95
2.1 – 5.0 lb	$18.95
5.1 – 10.0 lb	$20.95
10.1 – 20.0 lb	$22.95

What is the cost for overnight delivery of a package that weighs 8.5 pounds?
- Ⓐ $13.95
- Ⓑ $16.95
- Ⓒ $18.95
- Ⓓ $20.95

GO ON

Practice Test 2 *(continued)*

6. In which pair are the triangles congruent?

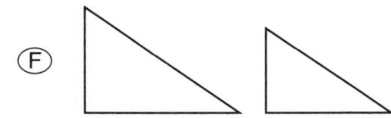

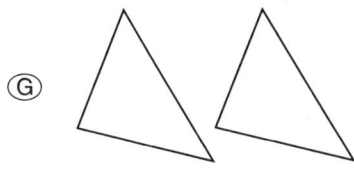

7. A trapezoid is a special quadrilateral in which only one pair of opposite sides are parallel. Which figure is a trapezoid?

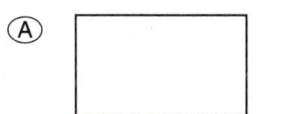

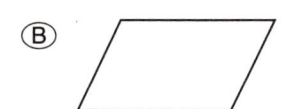

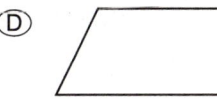

8. Which solid figure has five faces?
- Ⓕ sphere
- Ⓖ pyramid
- Ⓗ cube
- Ⓙ cone

9. What kind of angle is formed by the hands of a clock at 1:00?
- Ⓐ obtuse
- Ⓑ right
- Ⓒ acute
- Ⓓ straight

Use the picture to answer questions 10 and 11.

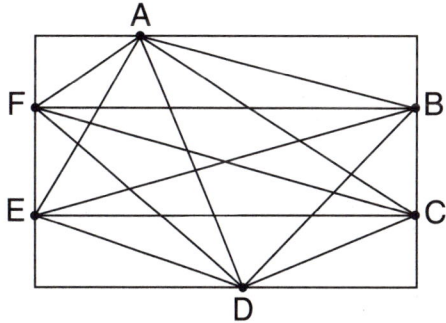

10. Which line segment is parallel to $\overline{FB}$?
- Ⓕ $\overline{AD}$ Ⓗ $\overline{EC}$
- Ⓖ $\overline{FD}$ Ⓙ $\overline{EB}$

11. Which is a right angle?
- Ⓐ ∠AFB Ⓒ ∠BCD
- Ⓑ ∠FEC Ⓓ ∠DEF

Practice Test 2 *(continued)*

12. Which figure shows figure WXYZ after a slide?

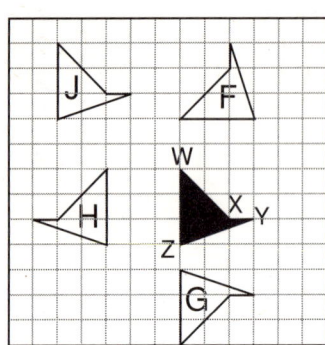

   Ⓕ Figure F   Ⓗ Figure H

   Ⓖ Figure G   Ⓙ Figure J

13. Which white figure shows where the shaded figure would be if the graph paper were folded along the dark line?

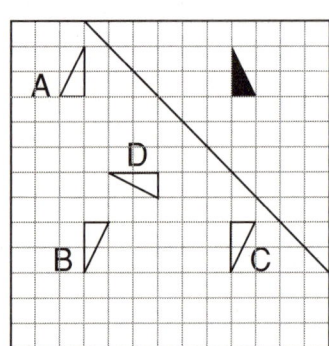

   Ⓐ Figure A   Ⓒ Figure C

   Ⓑ Figure B   Ⓓ Figure D

14. The rectangular yard at a dog kennel is 60 feet long and 35 feet wide. What is the perimeter of the yard?

   Ⓕ 25 ft   Ⓗ 190 ft

   Ⓖ 95 ft   Ⓙ 2100 ft

15. Each square in the figure below is a 1-centimeter square. What is the area of the figure?

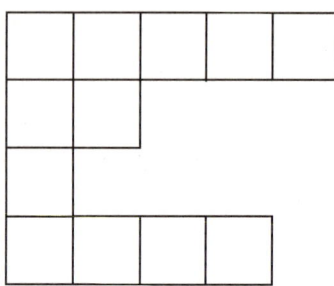

   Ⓐ 9 cm^2 Ⓒ 17 cm^2

   Ⓑ 12 cm^2 Ⓓ 20 cm^2

16. Find the area of a rectangular room that is 12 feet wide and 17 feet long.

   Ⓕ 204 sq ft   Ⓗ 51 sq ft

   Ⓖ 58 sq ft   Ⓙ 29 sq ft

17. This figure is made of 1-centimeter cubes. What is the volume of the figure?

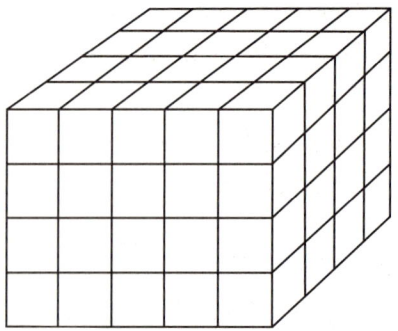

   Ⓐ 20 cm^3 Ⓒ 80 cm^3

   Ⓑ 56 cm^3 Ⓓ 320 cm^3

Practice Test 2 (continued)

18. Jan's plane took off at 11:35 A.M. Her watch showed this time when the plane landed.

How long was the flight?
- (F) 1 hour, 35 minutes
- (G) 2 hours, 25 minutes
- (H) 10 hours, 25 minutes
- (J) 10 hours, 35 minutes

19. Tim's clock stopped one afternoon at the time shown.

Later, at 6:05 P.M., he noticed that it had stopped. For how long had the clock been stopped?
- (A) 2 hours, 20 minutes
- (B) 3 hours, 45 minutes
- (C) 4 hours, 25 minutes
- (D) 8 hours, 15 minutes

20. Which unit should be used to measure the length of a driveway?
- (F) inches
- (G) gallons
- (H) miles
- (J) yards

21. Which item weighs about 1 kilogram?

(A)    (C)

(B)    (D)

22. What is the length of the caterpillar? (Use a centimeter ruler.)

- (F) 4.6 cm
- (G) 6 cm
- (H) 6.4 cm
- (J) 7.6 cm

23. What is the length of the paper clip? (Use an inch ruler.)

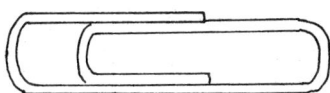

- (A) $1\frac{1}{2}$ inches
- (B) $1\frac{3}{4}$ inches
- (C) 2 inches
- (D) $2\frac{1}{4}$ inches

GO ON

Practice Test 2 *(continued)*

24. 75 centimeters is equivalent to —
  (F) 0.75 m
  (G) 7.5 m
  (H) 7500 mm
  (J) 0.075 km

25. Four men and 2 women step into an elevator. The combined weight of these adults is probably closest to —
  (A) 200 lb
  (B) 400 lb
  (C) 600 lb
  (D) 1000 lb

26. Use an inch ruler and the map below. What is the actual distance from Appleton to Pearsville?

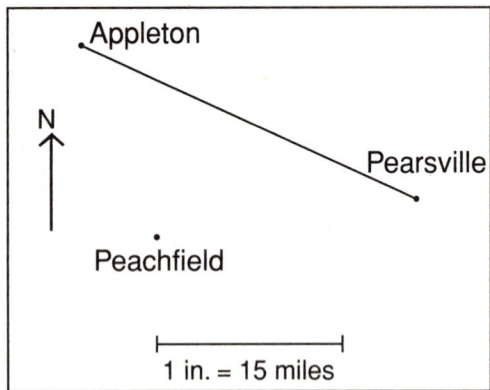

  (F) 75 miles
  (G) 30 miles
  (H) 15 miles
  (J) 2 miles

This bar graph shows the number of days students at Hillview Elementary School missed school each month last year. Use the graph to answer questions 27 and 28.

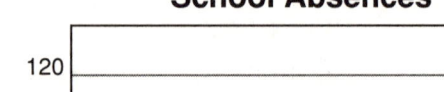

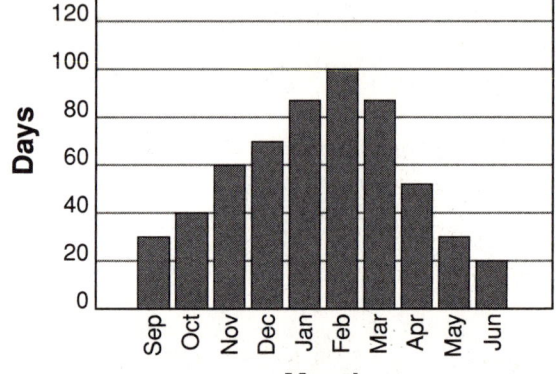

27. In which month were there the fewest absences?
  (A) February
  (B) May
  (C) June
  (D) September

28. About how many days did students miss school in December?
  (F) 40
  (G) 50
  (H) 60
  (J) 70

STOP

ANSWER SHEET

Student Name _____   Grade _____

Teacher Name _____   Date _____

MATHEMATICS

1 Ⓐ Ⓑ Ⓒ Ⓓ Ⓔ	11 Ⓐ Ⓑ Ⓒ Ⓓ Ⓔ	21 Ⓐ Ⓑ Ⓒ Ⓓ Ⓔ	31 Ⓐ Ⓑ Ⓒ Ⓓ Ⓔ
2 Ⓕ Ⓖ Ⓗ Ⓙ Ⓚ	12 Ⓕ Ⓖ Ⓗ Ⓙ Ⓚ	22 Ⓕ Ⓖ Ⓗ Ⓙ Ⓚ	32 Ⓕ Ⓖ Ⓗ Ⓙ Ⓚ
3 Ⓐ Ⓑ Ⓒ Ⓓ Ⓔ	13 Ⓐ Ⓑ Ⓒ Ⓓ Ⓔ	23 Ⓐ Ⓑ Ⓒ Ⓓ Ⓔ	33 Ⓐ Ⓑ Ⓒ Ⓓ Ⓔ
4 Ⓕ Ⓖ Ⓗ Ⓙ Ⓚ	14 Ⓕ Ⓖ Ⓗ Ⓙ Ⓚ	24 Ⓕ Ⓖ Ⓗ Ⓙ Ⓚ	34 Ⓕ Ⓖ Ⓗ Ⓙ Ⓚ
5 Ⓐ Ⓑ Ⓒ Ⓓ Ⓔ	15 Ⓐ Ⓑ Ⓒ Ⓓ Ⓔ	25 Ⓐ Ⓑ Ⓒ Ⓓ Ⓔ	35 Ⓐ Ⓑ Ⓒ Ⓓ Ⓔ
6 Ⓕ Ⓖ Ⓗ Ⓙ Ⓚ	16 Ⓕ Ⓖ Ⓗ Ⓙ Ⓚ	26 Ⓕ Ⓖ Ⓗ Ⓙ Ⓚ	36 Ⓕ Ⓖ Ⓗ Ⓙ Ⓚ
7 Ⓐ Ⓑ Ⓒ Ⓓ Ⓔ	17 Ⓐ Ⓑ Ⓒ Ⓓ Ⓔ	27 Ⓐ Ⓑ Ⓒ Ⓓ Ⓔ	37 Ⓐ Ⓑ Ⓒ Ⓓ Ⓔ
8 Ⓕ Ⓖ Ⓗ Ⓙ Ⓚ	18 Ⓕ Ⓖ Ⓗ Ⓙ Ⓚ	28 Ⓕ Ⓖ Ⓗ Ⓙ Ⓚ	38 Ⓕ Ⓖ Ⓗ Ⓙ Ⓚ
9 Ⓐ Ⓑ Ⓒ Ⓓ Ⓔ	19 Ⓐ Ⓑ Ⓒ Ⓓ Ⓔ	29 Ⓐ Ⓑ Ⓒ Ⓓ Ⓔ	39 Ⓐ Ⓑ Ⓒ Ⓓ Ⓔ
10 Ⓕ Ⓖ Ⓗ Ⓙ Ⓚ	20 Ⓕ Ⓖ Ⓗ Ⓙ Ⓚ	30 Ⓕ Ⓖ Ⓗ Ⓙ Ⓚ	40 Ⓕ Ⓖ Ⓗ Ⓙ Ⓚ

Practice
Test 3

Problem Solving

Name _____ Date _____

Practice Test 3

Directions. Choose the best answer to each question. Mark your answer. If the correct answer is *not given,* choose "NG."

1. Each bookshelf for a school book fair holds 12 books. How many shelves will be needed to display 75 books?
   - Ⓐ 6
   - Ⓑ 7
   - Ⓒ 63
   - Ⓓ 87
   - Ⓔ NG

2. There are 40 sweet wafers in a roll. A tin contains 23 rolls of wafers.

   How many wafers are in the tin?
   - Ⓕ 63 wafers
   - Ⓖ 92 wafers
   - Ⓗ 812 wafers
   - Ⓙ 920 wafers
   - Ⓚ NG

3. Tawana is making costumes for the school play. She has $4\frac{3}{8}$ yards of fabric. The pattern calls for $3\frac{1}{4}$ yards of fabric. How much fabric will be left over?
   - Ⓐ $1\frac{1}{8}$ yards
   - Ⓑ $1\frac{1}{4}$ yards
   - Ⓒ $1\frac{1}{2}$ yards
   - Ⓓ $7\frac{5}{8}$ yards
   - Ⓔ NG

4. Dave wants to buy a package of baseball cards that costs $3.75. If he saves $0.80 each week, how long will it take him to save enough money for the cards?
   - Ⓕ 4 weeks
   - Ⓖ 5 weeks
   - Ⓗ 6 weeks
   - Ⓙ 7 weeks
   - Ⓚ NG

5. Mr. Frederickson spent $22.10 on materials for a class art project. There are 34 students in the class. How much did Mr. Frederickson spend on each student?
   - Ⓐ $0.45
   - Ⓑ $0.50
   - Ⓒ $0.75
   - Ⓓ $11.90
   - Ⓔ NG

Practice Test 3 (continued)

6. The Social Committee spent $6.98 on paper goods, $9.32 on beverages, $13.04 on snacks, and $7 for decorations for the end-of-year party. How much did the committee spend in all?

- Ⓕ $36.34
- Ⓖ $35.24
- Ⓗ $29.41
- Ⓙ $17.71
- Ⓚ NG

7. Linda kept track of the time she spent on her science fair project over four days.

Day	Time Spent
Monday	43 minutes
Tuesday	1 hour, 10 min.
Wednesday	24 minutes
Thursday	$1\frac{1}{2}$ hours

How much time did Linda spend on her project all together?

- Ⓐ 2 hours, 47 minutes
- Ⓑ 3 hours, 7 minutes
- Ⓒ 3 hours, 27 minutes
- Ⓓ 3 hours, 47 minutes
- Ⓔ NG

8. Mr. Addison has 43 fence panels. Each one is 8 feet long. What is the longest fence Mr. Addison can build?

- Ⓕ 51 feet
- Ⓖ 172 feet
- Ⓗ 324 feet
- Ⓙ 344 feet
- Ⓚ NG

9. Jamaica started working in the garden at 10:45 A.M. She spent 45 minutes weeding, $1\frac{1}{2}$ hours mowing, and 25 minutes watering. What time did she finish?

- Ⓐ 12:25 P.M.
- Ⓑ 12:40 P.M.
- Ⓒ 1:00 P.M.
- Ⓓ 1:25 P.M.
- Ⓔ NG

10. There are 20,324 students in a school district. Of them, 1821 students get to school by walking or by bicycle. The rest take the school buses. <u>About</u> how many students take the buses?

- Ⓕ 18,000
- Ⓖ 19,000
- Ⓗ 20,000
- Ⓙ 22,000

11. A car powered by electricity and gasoline travels 68 miles per gallon of gasoline. Its tank holds 10.6 gallons of gas. <u>About</u> how far can the car go on a full tank of gas?

- Ⓐ 7 miles
- Ⓑ 70 miles
- Ⓒ 700 miles
- Ⓓ 7000 miles

12. The average class size in one school district is 32 students. If there are 987 classes in the district, <u>about</u> how many students are there?

- Ⓕ 40,000
- Ⓖ 30,000
- Ⓗ 27,000
- Ⓙ 25,000

GO ON

Scholastic Professional Books

Practice Test 3 (continued)

13. The 20 members of the soccer team voted on a new mascot. Three fifths of them voted for the gazelle. How many votes were for the gazelle?

- Ⓐ 4 votes
- Ⓑ 12 votes
- Ⓒ 15 votes
- Ⓓ 33 votes
- Ⓔ NG

14. Peter's family drove 171 miles in 3 hours. At the same rate, how far will they travel in 10 hours?

- Ⓕ 399 miles
- Ⓖ 480 miles
- Ⓗ 520 miles
- Ⓙ 640 miles
- Ⓚ NG

15. A map has a scale in which 2.5 cm represents 15 miles. What does 4 cm represent on the map?

- Ⓐ 6 miles
- Ⓑ 10 miles
- Ⓒ 24 miles
- Ⓓ 30 miles
- Ⓔ NG

Use the chart for questions 16 and 17.

Kendra designed a race track for marbles. She pours her marbles through a funnel onto the track.

Color	Number of Marbles
Green	5
Blue	7
Red	2
Purple	6

16. Which color marble is most likely to win?

- Ⓕ green
- Ⓖ blue
- Ⓗ red
- Ⓙ purple
- Ⓚ NG

17. What is the probability of a red marble finishing first?

- Ⓐ $\frac{1}{5}$
- Ⓑ $\frac{1}{10}$
- Ⓒ $\frac{1}{4}$
- Ⓓ $\frac{1}{8}$
- Ⓔ NG

18. A total of 300 tickets for a door prize were passed out. Mr. Ryan got 6 tickets. What is the probability that one of Mr. Ryan's tickets will be drawn for the prize?

- Ⓕ $\frac{50}{1}$
- Ⓖ $\frac{3}{25}$
- Ⓗ $\frac{1}{50}$
- Ⓙ $\frac{1}{300}$
- Ⓚ NG

Practice Test 3 *(continued)*

19. Matthew has a bag with 9 coins. He has twice as many dimes as quarters. He has the same number of quarters and nickels. He has one penny. Each coin is worth $0.25 or less. What is the total value of Matthew's coins?

Ⓐ $0.51
Ⓑ $0.81
Ⓒ $0.96
Ⓓ $1.01
Ⓔ NG

20. Five people are standing in a line at the video store. Pete is at the front of the line. Stan is directly in front of Brad. Alison is between Stan and Barb. Who is last in line?

Ⓕ Alison
Ⓖ Stan
Ⓗ Barb
Ⓙ Brad
Ⓚ NG

21. Jed is planning his weekly radio show. There are 7 minutes left in the program. Along with the first track of the featured CD, which track should Jed play to fill the time without going over 7 minutes?

Ⓐ Track 2
Ⓑ Track 3
Ⓒ Track 4
Ⓓ Track 5
Ⓔ NG

Track	Time
1	4:21
2	1:05
3	2:35
4	3:06
5	2:55
6	3:15

22. It takes Julie $1\frac{2}{3}$ hours to make a gecko with beads. Which number sentence could be used to find how many geckos Julie can make in 15 hours?

Ⓕ $15 \div 1\frac{2}{3} = \square$
Ⓖ $15 + 1\frac{2}{3} = \square$
Ⓗ $15 \times 1\frac{2}{3} = \square$
Ⓙ $15 - 1\frac{2}{3} = \square$
Ⓚ NG

23. Kara plans to paint a wall that is 20 feet long and 9 feet high. There is a 3 ft. by 7 ft. door in the wall. Which number sentence could be used to find the area of the wall to be painted?

Ⓐ $(20 \times 9) - (3 \times 7) = \square$
Ⓑ $(20 \times 9) - 3 - 7 = \square$
Ⓒ $20 \times 9 = \square$
Ⓓ $(20 \times 9) + (3 \times 7) = \square$
Ⓔ NG

24. Ralph bicycled 15 miles on Friday, 22 miles on Saturday, and 17 miles on Sunday. Which number sentence could be used to find the average number of miles he bicycled each day?

Ⓕ $15 + 22 + 17 = \square$
Ⓖ $22 - 17 = \square$
Ⓗ $(15 + 22 + 17) \times 3 = \square$
Ⓙ $(15 + 22 + 17) - 3 = \square$
Ⓚ NG

GO ON ▷

Practice Test 3 (continued)

25. Olivia put $35.75 in a new savings account. The bank pays 6% interest each year. What information do you need to figure out how much money is in Olivia's account now?

ⓐ how long the money has been in the account

ⓑ the type of bank she used

ⓒ the name of the bank

ⓓ the savings account number

ⓔ NG

26. When Mrs. Heath's fifth-grade students raised money for a class trip, 24% of the money was earned by washing cars. What information do you need to figure out how much was raised at the car wash?

ⓕ the cost per car for the wash

ⓖ the number of cars washed

ⓗ the number of students

ⓙ the total amount of money raised

ⓚ NG

27. Mrs. Gulledge had a box of 72 pencils. She gave two pencils to each fifth grader. What else do you need to know to find out how many pencils are left over?

ⓐ how many pencils each student has

ⓑ how often Mrs. Gulledge gives out pencils

ⓒ how much the pencils cost

ⓓ how many students are in the class

ⓔ NG

28. A camel can hold 22 gallons of water in its stomachs. Each pint of water weighs 1 pound. How many pounds of water can the camel hold?

ⓕ 44 lb

ⓖ 88 lb

ⓗ 176 lb

ⓙ 352 lb

ⓚ NG

29. Mario wants to order a shirt from a catalog. The price of the shirt is $23. The state sales tax is 5%. The cost for shipping is $7.95. What is the total cost for the shirt?

ⓐ $33.10

ⓑ $30.95

ⓒ $29.10

ⓓ $24.15

ⓔ NG

30. After school, Randy spent $2\frac{1}{2}$ hours at soccer practice, 25 minutes on spelling homework, $\frac{3}{4}$ hour on math problems, and $1\frac{1}{2}$ hours on a social studies project. How much more time did Randy spend on homework than on soccer?

ⓕ 5 hours, 10 minutes

ⓖ 2 hours, 40 minutes

ⓗ 2 hours, 20 minutes

ⓙ 10 minutes

ⓚ NG

STOP

ANSWER SHEET

Practice Test #3

Student Name _____

Grade _____

Teacher Name _____

Date _____

MATHEMATICS

1 Ⓐ Ⓑ Ⓒ Ⓓ Ⓔ	**11** Ⓐ Ⓑ Ⓒ Ⓓ Ⓔ	**21** Ⓐ Ⓑ Ⓒ Ⓓ Ⓔ	**31** Ⓐ Ⓑ Ⓒ Ⓓ Ⓔ
2 Ⓕ Ⓖ Ⓗ Ⓙ Ⓚ	**12** Ⓕ Ⓖ Ⓗ Ⓙ Ⓚ	**22** Ⓕ Ⓖ Ⓗ Ⓙ Ⓚ	**32** Ⓕ Ⓖ Ⓗ Ⓙ Ⓚ
3 Ⓐ Ⓑ Ⓒ Ⓓ Ⓔ	**13** Ⓐ Ⓑ Ⓒ Ⓓ Ⓔ	**23** Ⓐ Ⓑ Ⓒ Ⓓ Ⓔ	**33** Ⓐ Ⓑ Ⓒ Ⓓ Ⓔ
4 Ⓕ Ⓖ Ⓗ Ⓙ Ⓚ	**14** Ⓕ Ⓖ Ⓗ Ⓙ Ⓚ	**24** Ⓕ Ⓖ Ⓗ Ⓙ Ⓚ	**34** Ⓕ Ⓖ Ⓗ Ⓙ Ⓚ
5 Ⓐ Ⓑ Ⓒ Ⓓ Ⓔ	**15** Ⓐ Ⓑ Ⓒ Ⓓ Ⓔ	**25** Ⓐ Ⓑ Ⓒ Ⓓ Ⓔ	**35** Ⓐ Ⓑ Ⓒ Ⓓ Ⓔ
6 Ⓕ Ⓖ Ⓗ Ⓙ Ⓚ	**16** Ⓕ Ⓖ Ⓗ Ⓙ Ⓚ	**26** Ⓕ Ⓖ Ⓗ Ⓙ Ⓚ	**36** Ⓕ Ⓖ Ⓗ Ⓙ Ⓚ
7 Ⓐ Ⓑ Ⓒ Ⓓ Ⓔ	**17** Ⓐ Ⓑ Ⓒ Ⓓ Ⓔ	**27** Ⓐ Ⓑ Ⓒ Ⓓ Ⓔ	**37** Ⓐ Ⓑ Ⓒ Ⓓ Ⓔ
8 Ⓕ Ⓖ Ⓗ Ⓙ Ⓚ	**18** Ⓕ Ⓖ Ⓗ Ⓙ Ⓚ	**28** Ⓕ Ⓖ Ⓗ Ⓙ Ⓚ	**38** Ⓕ Ⓖ Ⓗ Ⓙ Ⓚ
9 Ⓐ Ⓑ Ⓒ Ⓓ Ⓔ	**19** Ⓐ Ⓑ Ⓒ Ⓓ Ⓔ	**29** Ⓐ Ⓑ Ⓒ Ⓓ Ⓔ	**39** Ⓐ Ⓑ Ⓒ Ⓓ Ⓔ
10 Ⓕ Ⓖ Ⓗ Ⓙ Ⓚ	**20** Ⓕ Ⓖ Ⓗ Ⓙ Ⓚ	**30** Ⓕ Ⓖ Ⓗ Ⓙ Ⓚ	**40** Ⓕ Ⓖ Ⓗ Ⓙ Ⓚ

Practice
Test 4
Computation

Practice Test 4

Directions. Choose the best answer to each question. Mark your answer. If the correct answer is *not given,* choose "NG."

1.
$$\begin{array}{r} 40 \\ \times\ 27 \end{array}$$

Ⓐ 360
Ⓑ 828
Ⓒ 1010
Ⓓ 1080
Ⓔ NG

2. 6)56

Ⓕ 9 R2
Ⓖ 9 R1
Ⓗ 8 R4
Ⓙ 8 R2
Ⓚ NG

3. 15)620

Ⓐ 42 R5
Ⓑ 41 R5
Ⓒ 41 R3
Ⓓ 40 R5
Ⓔ NG

4. This chart shows the number of movie tickets sold at the theater each day.

Movie Tickets Sold	
Monday	12
Tuesday	20
Wednesday	28
Thursday	36
Friday	54

What was the average number of tickets sold per day?
Ⓕ 150
Ⓖ 37
Ⓗ 30
Ⓙ 28
Ⓚ NG

5. Amy bought 8 boxes of tiles.

TILES
TOTAL COUNT: 108

How many tiles did she buy in all?
Ⓐ 116
Ⓑ 816
Ⓒ 860
Ⓓ 1864
Ⓔ NG

GO ON

Practice Test 4 (continued)

6.
$$\frac{3}{5}$$
$$+\frac{4}{5}$$

Ⓕ $\frac{7}{25}$

Ⓖ $1\frac{2}{5}$

Ⓗ $\frac{5}{7}$

Ⓙ $\frac{12}{25}$

Ⓚ NG

7.
$$2\frac{1}{3}$$
$$+1\frac{1}{2}$$

Ⓐ $3\frac{2}{5}$

Ⓑ $3\frac{1}{3}$

Ⓒ $3\frac{5}{6}$

Ⓓ $4\frac{1}{6}$

Ⓔ NG

8. $\frac{5}{9} - \frac{2}{9} =$

Ⓕ $\frac{7}{9}$

Ⓖ $\frac{10}{18}$

Ⓗ $\frac{3}{18}$

Ⓙ $\frac{2}{3}$

Ⓚ NG

9. $\frac{3}{4} - \frac{1}{3} =$

Ⓐ $\frac{5}{12}$

Ⓑ $\frac{2}{1}$

Ⓒ $\frac{3}{7}$

Ⓓ $\frac{1}{4}$

Ⓔ NG

10. A restaurant menu offers 4 kinds of sandwiches, 3 kinds of salads, and 2 kinds of drinks.

Menu		
Sandwiches 4	Salads 3	Drinks 2

How many different combinations of 1 sandwich, 1 salad, and 1 drink can be made?

Ⓕ 9

Ⓖ 12

Ⓗ 20

Ⓙ 24

Ⓚ NG

11. Lara is playing a game with letter tiles. These tiles are in a box.

Letter	Number of Tiles
E	12
T	6
A	4
R	8

If Lara takes out one tile without looking, what is the probability that she will get a "T"?

Ⓐ $\frac{4}{5}$

Ⓑ $\frac{1}{4}$

Ⓒ $\frac{1}{30}$

Ⓓ $\frac{1}{5}$

Ⓔ NG

Scholastic Professional Books

Practice Test 4 *(continued)*

12. $\dfrac{4}{5} \times \dfrac{2}{3} =$

    Ⓕ $\dfrac{8}{15}$

    Ⓖ $\dfrac{6}{8}$

    Ⓗ $\dfrac{6}{15}$

    Ⓙ $\dfrac{8}{8}$

    Ⓚ NG

13. $\dfrac{3}{10} \times \dfrac{1}{2} =$

    Ⓐ $\dfrac{3}{12}$

    Ⓑ $\dfrac{4}{12}$

    Ⓒ $\dfrac{3}{20}$

    Ⓓ $\dfrac{4}{20}$

    Ⓔ NG

14. $15.95
    $+\ 6.28$

    Ⓕ $20.23
    Ⓖ $21.13
    Ⓗ $22.13
    Ⓙ $22.23
    Ⓚ NG

15. $1.84 + 2.3 =$

    Ⓐ 3.14
    Ⓑ 3.87
    Ⓒ 4.04
    Ⓓ 4.24
    Ⓔ NG

Use the grid map below to answer questions 16 and 17.

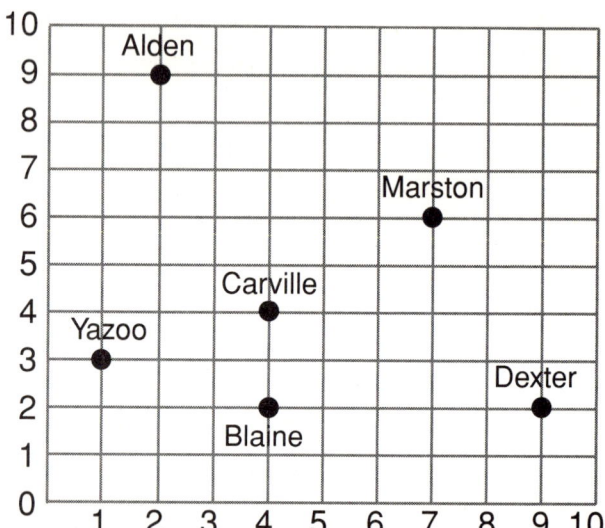

16. What city is located at (4, 2)?

    Ⓕ Blaine
    Ⓖ Yazoo
    Ⓗ Carville
    Ⓙ Dexter
    Ⓚ NG

17. Where is Marston located?

    Ⓐ (9, 2)
    Ⓑ (7, 6)
    Ⓒ (4, 4)
    Ⓓ (6, 7)
    Ⓔ NG

GO ON ⇨

Practice Test 4 *(continued)*

18. $6.3 - 0.8 =$

  Ⓕ 7.1
  Ⓖ 6.5
  Ⓗ 5.6
  Ⓙ 4.5
  Ⓚ NG

19. $\$20 - \$8.95 =$

  Ⓐ $12.05
  Ⓑ $11.05
  Ⓒ $10.95
  Ⓓ $10.50
  Ⓔ NG

20. $3.2 \times 0.5 =$

  Ⓕ 0.3
  Ⓖ 1.5
  Ⓗ 1.6
  Ⓙ 16
  Ⓚ NG

21. $25 \times \$2.40 =$

  Ⓐ $6.00
  Ⓑ $15.00
  Ⓒ $51.00
  Ⓓ $60.00
  Ⓔ NG

22. This chart shows the number of miles Trevor rode his bike each day.

Miles Trevor Rode	
Monday	45
Tuesday	40
Wednesday	32
Thursday	38
Friday	35

What was the average number of miles he rode per day?

  Ⓕ 28
  Ⓖ 32
  Ⓗ 38
  Ⓙ 190
  Ⓚ NG

23. Marlo wrote this number sentence to solve a problem.

$(6 \times 2n) \div 8 = 15$

What is the value of *n*?

  Ⓐ 8
  Ⓑ 9
  Ⓒ 10
  Ⓓ 15
  Ⓔ NG

GO ON ⇨

Practice Test 4 (continued)

24. Franny rolls two number cubes numbered 1–6. What is the probability that she will roll 12 on one try?

- Ⓕ $\frac{1}{36}$
- Ⓖ $\frac{1}{18}$
- Ⓗ $\frac{1}{30}$
- Ⓙ $\frac{12}{35}$
- Ⓚ NG

25. If $(3x + 5x) - 4 = 36$, then what is the value of x?

- Ⓐ 8
- Ⓑ 5
- Ⓒ 4
- Ⓓ 3
- Ⓔ NG

26. Collin wrote this equation to solve a problem.

$$7y - 5 = 58$$

What is the value of y?

- Ⓕ 5
- Ⓖ 6
- Ⓗ 7
- Ⓙ 8
- Ⓚ NG

Use the graph below to answer questions 27 and 28.

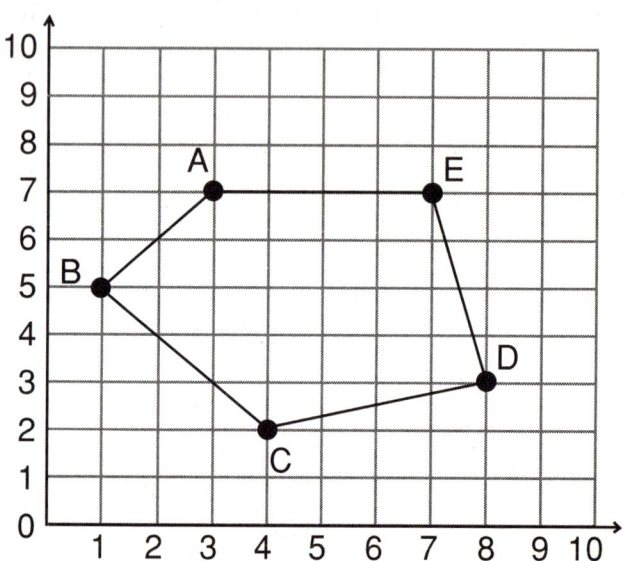

27. What are the coordinates for the location of point D?

- Ⓐ (3, 8)
- Ⓑ (4, 2)
- Ⓒ (4, 4)
- Ⓓ (8, 3)
- Ⓔ NG

28. Which point is located at (3, 7)?

- Ⓕ point A
- Ⓖ point B
- Ⓗ point C
- Ⓙ point E
- Ⓚ NG

STOP

ANSWER SHEET

Student Name _____ Grade _____

Teacher Name _____ Date _____

MATHEMATICS

1 Ⓐ Ⓑ Ⓒ Ⓓ Ⓔ	11 Ⓐ Ⓑ Ⓒ Ⓓ Ⓔ	21 Ⓐ Ⓑ Ⓒ Ⓓ Ⓔ	31 Ⓐ Ⓑ Ⓒ Ⓓ Ⓔ
2 Ⓕ Ⓖ Ⓗ Ⓙ Ⓚ	12 Ⓕ Ⓖ Ⓗ Ⓙ Ⓚ	22 Ⓕ Ⓖ Ⓗ Ⓙ Ⓚ	32 Ⓕ Ⓖ Ⓗ Ⓙ Ⓚ
3 Ⓐ Ⓑ Ⓒ Ⓓ Ⓔ	13 Ⓐ Ⓑ Ⓒ Ⓓ Ⓔ	23 Ⓐ Ⓑ Ⓒ Ⓓ Ⓔ	33 Ⓐ Ⓑ Ⓒ Ⓓ Ⓔ
4 Ⓕ Ⓖ Ⓗ Ⓙ Ⓚ	14 Ⓕ Ⓖ Ⓗ Ⓙ Ⓚ	24 Ⓕ Ⓖ Ⓗ Ⓙ Ⓚ	34 Ⓕ Ⓖ Ⓗ Ⓙ Ⓚ
5 Ⓐ Ⓑ Ⓒ Ⓓ Ⓔ	15 Ⓐ Ⓑ Ⓒ Ⓓ Ⓔ	25 Ⓐ Ⓑ Ⓒ Ⓓ Ⓔ	35 Ⓐ Ⓑ Ⓒ Ⓓ Ⓔ
6 Ⓕ Ⓖ Ⓗ Ⓙ Ⓚ	16 Ⓕ Ⓖ Ⓗ Ⓙ Ⓚ	26 Ⓕ Ⓖ Ⓗ Ⓙ Ⓚ	36 Ⓕ Ⓖ Ⓗ Ⓙ Ⓚ
7 Ⓐ Ⓑ Ⓒ Ⓓ Ⓔ	17 Ⓐ Ⓑ Ⓒ Ⓓ Ⓔ	27 Ⓐ Ⓑ Ⓒ Ⓓ Ⓔ	37 Ⓐ Ⓑ Ⓒ Ⓓ Ⓔ
8 Ⓕ Ⓖ Ⓗ Ⓙ Ⓚ	18 Ⓕ Ⓖ Ⓗ Ⓙ Ⓚ	28 Ⓕ Ⓖ Ⓗ Ⓙ Ⓚ	38 Ⓕ Ⓖ Ⓗ Ⓙ Ⓚ
9 Ⓐ Ⓑ Ⓒ Ⓓ Ⓔ	19 Ⓐ Ⓑ Ⓒ Ⓓ Ⓔ	29 Ⓐ Ⓑ Ⓒ Ⓓ Ⓔ	39 Ⓐ Ⓑ Ⓒ Ⓓ Ⓔ
10 Ⓕ Ⓖ Ⓗ Ⓙ Ⓚ	20 Ⓕ Ⓖ Ⓗ Ⓙ Ⓚ	30 Ⓕ Ⓖ Ⓗ Ⓙ Ⓚ	40 Ⓕ Ⓖ Ⓗ Ⓙ Ⓚ

Practice Test 5

Numeration and Number Concepts

Name _____ Date _____

Practice Test 5

Directions. Choose the best answer to each question. Mark your answer.

1. **Brazil has an area of about 3,287,000 square miles. How is 3,287,000 written in words?**
   Ⓐ thirty-two million eight hundred seventy thousand
   Ⓑ three million two hundred eighty-seven thousand
   Ⓒ three hundred twenty-eight thousand seven hundred
   Ⓓ three million two hundred thousand eighty-seven

2. **The population of Belize is about two hundred forty-nine thousand three hundred ten. How is this number written as a numeral?**
   Ⓕ 200,049,310
   Ⓖ 204,931
   Ⓗ 249,310
   Ⓙ 249,301

3. **Which is an odd number?**
   Ⓐ 2060
   Ⓑ 7254
   Ⓒ 1102
   Ⓓ 3481

4. **The chart shows the population of four states in 2000.**

State	Population
Indiana	6,080,485
Missouri	5,595,211
Tennessee	5,689,283
Washington	5,894,121

   Which state has the smallest population?
   Ⓕ Indiana
   Ⓖ Missouri
   Ⓗ Tennessee
   Ⓙ Washington

5. **The chart shows the results of the 2000 presidential election in Pennsylvania.**

Candidate	Number of Votes
Buchanan	16,879
Bush	2,264,309
Gore	2,465,412
Nader	102,453

   Which list shows the candidates in order from the greatest number of votes received to the least?
   Ⓐ Gore, Bush, Nader, Buchanan
   Ⓑ Bush, Gore, Buchanan, Nader
   Ⓒ Bush, Gore, Nader, Buchanan
   Ⓓ Gore, Bush, Buchanan, Nader

GO ON

Practice Test 5 (continued)

6. At its farthest point, the moon is 251,966 miles from Earth. What is that number rounded to the nearest ten thousand miles?

(F) 251,970
(G) 251,900
(H) 252,000
(J) 250,000

7. This sign shows the number of cars sold by a car dealer.

1	8	7	9	3	4	5

The "9" represents what place value in this number?

(A) hundreds
(B) thousands
(C) ten thousands
(D) hundred thousands

8. $70,000 + 400 + 80 =$

(F) 70,480
(G) 700,480
(H) 70,048
(J) 7480

9. Last year, Darcy earned $14,685.00. What is that number rounded to the nearest hundred dollars?

(A) $15,000
(B) $14,600
(C) $14,690
(D) $14,700

10. Drake created this number pattern.

0, 6, 13, 21, 30, 40, _____

If the pattern continues, what number should go in the blank?

(F) 49
(G) 50
(H) 51
(J) 52

11. An artist designed this carpet pattern.

If the pattern continues, what design would fit in the empty block?

(A)   (C)

(B) 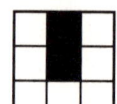   (D)

GO ON

Practice Test 5 *(continued)*

12. Which of these will result in an even number?

 (F) $5 + 7 + 9$
 (G) $8 + 0$
 (H) $5 - 3 - 1$
 (J) $9 - 0$

13. The list shows the number of people from three states who visited Niagara Falls.

Pennsylvania	6134
Michigan	4288
Ohio	3990

<u>About</u> how many visitors were there all together from these three states?

 (A) 12,500
 (B) 13,000
 (C) 13,500
 (D) 14,000

14. In a large banquet hall, there are 38 tables. Each table seats 22 people. <u>About</u> how many people can be seated in the hall at one time?

 (F) 8000
 (G) 6000
 (H) 800
 (J) 600

15. What is the greatest common factor of 9, 12, and 27?

 (A) 2
 (B) 3
 (C) 6
 (D) 9

16. What is the least common multiple of 8 and 20?

 (F) 4
 (G) 40
 (H) 80
 (J) 160

17. Look at the number line.

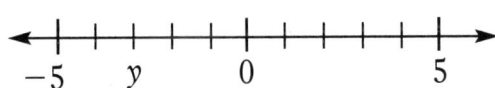

What number belongs in the place marked with *y?*

 (A) 3
 (B) -1
 (C) -2
 (D) -3

GO ON ⇒

Practice Test 5 *(continued)*

18. There are 24 students in Ms. Granger's class, and 12 of them are girls. What fraction of the students are girls?

- Ⓕ $\frac{1}{2}$
- Ⓖ $\frac{1}{3}$
- Ⓗ $\frac{3}{5}$
- Ⓙ $\frac{2}{3}$

19. Of the 100 horses in a corral, 75 are brown. What fraction of the horses are brown?

- Ⓐ $\frac{7}{10}$
- Ⓑ $\frac{4}{5}$
- Ⓒ $\frac{3}{4}$
- Ⓓ $\frac{2}{3}$

20. Which number line shows the difference of $3 - 5$?

Ⓕ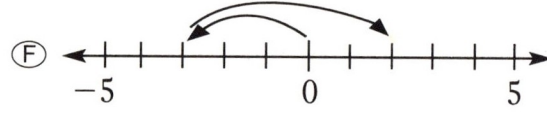

Ⓖ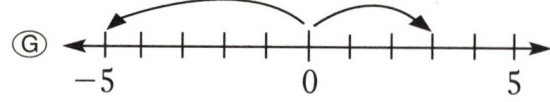

Ⓗ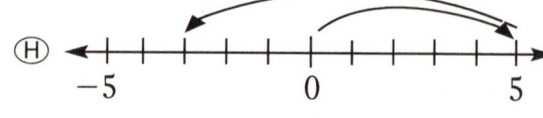

Ⓙ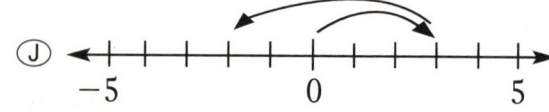

21. Which bottle of juice holds the most?

Brand A	$1\frac{1}{2}$ qt
Brand B	$1\frac{3}{4}$ qt
Brand C	$1\frac{2}{3}$ qt
Brand D	$1\frac{5}{6}$ qt

- Ⓐ Brand A
- Ⓑ Brand B
- Ⓒ Brand C
- Ⓓ Brand D

22. In which list are the fractions arranged from greatest to least?

- Ⓕ $\frac{3}{10}, \frac{3}{5}, \frac{1}{2}, \frac{1}{4}$
- Ⓖ $\frac{3}{5}, \frac{1}{2}, \frac{3}{10}, \frac{1}{4}$
- Ⓗ $\frac{1}{4}, \frac{1}{2}, \frac{3}{10}, \frac{3}{5}$
- Ⓙ $\frac{1}{2}, \frac{1}{4}, \frac{3}{5}, \frac{3}{10}$

23. An oil tank holds 250 gallons. It has 100 gallons of oil in it. What fraction tells how full the tank is?

- Ⓐ $\frac{1}{2}$
- Ⓑ $\frac{2}{5}$
- Ⓒ $\frac{3}{5}$
- Ⓓ $\frac{2}{3}$

GO ON ⟩

Scholastic Professional Books

Practice Test 5 *(continued)*

24. Which decimal number is equal to $\frac{1}{4}$?

  Ⓕ 0.20

  Ⓖ 0.25

  Ⓗ 0.45

  Ⓙ 0.50

25. In a race, four swimmers finished with these times.

James	30.42
Rosa	30.65
Lily	30.33
Rafael	30.71

Which list shows the swimmers in order from fastest time to slowest?

  Ⓐ Lily, James, Rosa, Rafael

  Ⓑ Rafael, Rosa, James, Lily

  Ⓒ Lily, Rosa, Rafael, James

  Ⓓ James, Rosa, Rafael, Lily

26. At the grocery store, Mara found four loaves of bread at four different prices. Which is the highest price?

  Ⓕ  $2.48

  Ⓖ  $2.55

  Ⓗ  $2.69

  Ⓙ  $2.63

27. What number is shown on the number line?

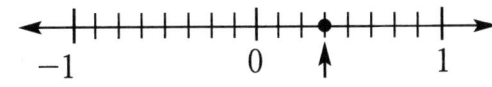

  Ⓐ $\frac{1}{2}$ Ⓒ $\frac{1}{4}$

  Ⓑ $\frac{3}{8}$ Ⓓ $\frac{3}{5}$

28. Tino found $0.60. What fraction of a dollar did he find?

  Ⓕ $\frac{2}{3}$ Ⓗ $\frac{3}{4}$

  Ⓖ $\frac{1}{2}$ Ⓙ $\frac{3}{5}$

29. $(3x + 1) \times 0 =$

  Ⓐ $3x + 1$

  Ⓑ $3x + 10$

  Ⓒ 0

  Ⓓ $1 + 0$

30. Which statement is true?

  Ⓕ $9 \times \frac{1}{9} = 1$

  Ⓖ $\frac{1}{2} \times \frac{1}{2} = \frac{1}{2} + \frac{1}{2}$

  Ⓗ $3 \times \frac{1}{3} = 3$

  Ⓙ $\frac{1}{4} \times \frac{1}{4} = \frac{1}{8}$

STOP

ANSWER SHEET

Student Name _____ Grade _____

Teacher Name _____ Date _____

MATHEMATICS

1 Ⓐ Ⓑ Ⓒ Ⓓ Ⓔ	11 Ⓐ Ⓑ Ⓒ Ⓓ Ⓔ	21 Ⓐ Ⓑ Ⓒ Ⓓ Ⓔ	31 Ⓐ Ⓑ Ⓒ Ⓓ Ⓔ
2 Ⓕ Ⓖ Ⓗ Ⓙ Ⓚ	12 Ⓕ Ⓖ Ⓗ Ⓙ Ⓚ	22 Ⓕ Ⓖ Ⓗ Ⓙ Ⓚ	32 Ⓕ Ⓖ Ⓗ Ⓙ Ⓚ
3 Ⓐ Ⓑ Ⓒ Ⓓ Ⓔ	13 Ⓐ Ⓑ Ⓒ Ⓓ Ⓔ	23 Ⓐ Ⓑ Ⓒ Ⓓ Ⓔ	33 Ⓐ Ⓑ Ⓒ Ⓓ Ⓔ
4 Ⓕ Ⓖ Ⓗ Ⓙ Ⓚ	14 Ⓕ Ⓖ Ⓗ Ⓙ Ⓚ	24 Ⓕ Ⓖ Ⓗ Ⓙ Ⓚ	34 Ⓕ Ⓖ Ⓗ Ⓙ Ⓚ
5 Ⓐ Ⓑ Ⓒ Ⓓ Ⓔ	15 Ⓐ Ⓑ Ⓒ Ⓓ Ⓔ	25 Ⓐ Ⓑ Ⓒ Ⓓ Ⓔ	35 Ⓐ Ⓑ Ⓒ Ⓓ Ⓔ
6 Ⓕ Ⓖ Ⓗ Ⓙ Ⓚ	16 Ⓕ Ⓖ Ⓗ Ⓙ Ⓚ	26 Ⓕ Ⓖ Ⓗ Ⓙ Ⓚ	36 Ⓕ Ⓖ Ⓗ Ⓙ Ⓚ
7 Ⓐ Ⓑ Ⓒ Ⓓ Ⓔ	17 Ⓐ Ⓑ Ⓒ Ⓓ Ⓔ	27 Ⓐ Ⓑ Ⓒ Ⓓ Ⓔ	37 Ⓐ Ⓑ Ⓒ Ⓓ Ⓔ
8 Ⓕ Ⓖ Ⓗ Ⓙ Ⓚ	18 Ⓕ Ⓖ Ⓗ Ⓙ Ⓚ	28 Ⓕ Ⓖ Ⓗ Ⓙ Ⓚ	38 Ⓕ Ⓖ Ⓗ Ⓙ Ⓚ
9 Ⓐ Ⓑ Ⓒ Ⓓ Ⓔ	19 Ⓐ Ⓑ Ⓒ Ⓓ Ⓔ	29 Ⓐ Ⓑ Ⓒ Ⓓ Ⓔ	39 Ⓐ Ⓑ Ⓒ Ⓓ Ⓔ
10 Ⓕ Ⓖ Ⓗ Ⓙ Ⓚ	20 Ⓕ Ⓖ Ⓗ Ⓙ Ⓚ	30 Ⓕ Ⓖ Ⓗ Ⓙ Ⓚ	40 Ⓕ Ⓖ Ⓗ Ⓙ Ⓚ

Practice Test 6

Geometry and Measurement

Practice Test 6

Directions. Choose the best answer to each question. Mark your answer.

1. Which figure is a pentagon?

2. Merrill rode 3.5 kilometers on his bike. How many meters is that?

- Ⓕ 0.35
- Ⓖ 35
- Ⓗ 350
- Ⓙ 3500

3. Which figure has 5 faces?

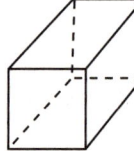

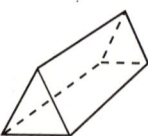

4. Which figure shows a line of symmetry?

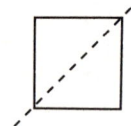   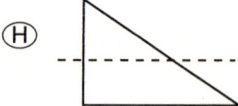

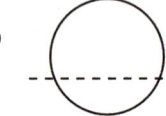   

5. Suppose that this figure is turned one 90-degree turn in the direction of the arrow.

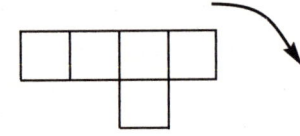

Which picture shows how the figure would look after it has been turned?

Ⓐ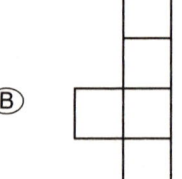

Ⓑ

Ⓒ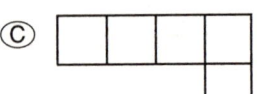

Ⓓ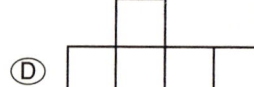

GO ON

Scholastic Professional Books

Practice Test 6 *(continued)*

6. Look at Figure A

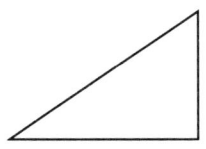

Fig. A

Which is congruent to Figure A?

 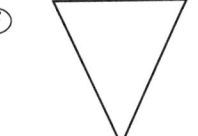   Ⓗ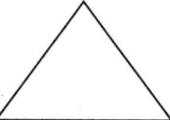

Ⓖ 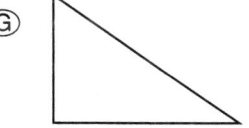   Ⓙ

7. There is a rectangular basketball court at the city playground.

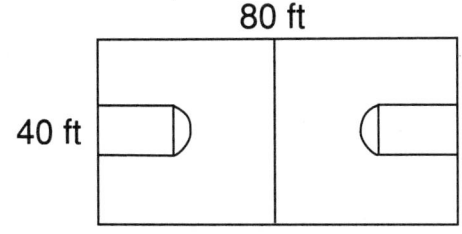

What is the perimeter of the court?
Ⓐ 120 ft
Ⓑ 200 ft
Ⓒ 240 ft
Ⓓ 3200 ft

Use the picture to answer questions 8 and 9.

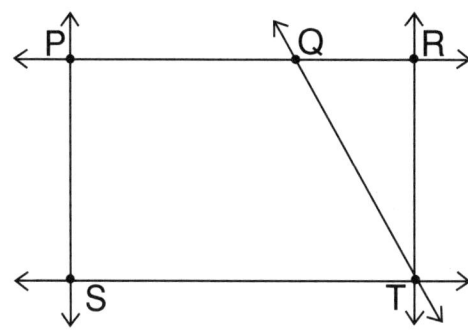

8. Which line is parallel to $\overleftrightarrow{PR}$?
Ⓕ $\overleftrightarrow{PS}$
Ⓖ $\overleftrightarrow{QT}$
Ⓗ $\overleftrightarrow{RT}$
Ⓙ $\overleftrightarrow{ST}$

9. Which is a right angle?
Ⓐ ∠TSP
Ⓑ ∠PQT
Ⓒ ∠QTR
Ⓓ ∠RQT

10. What kind of angle is formed by the hands on the clock?

Ⓕ straight
Ⓖ right
Ⓗ acute
Ⓙ obtuse

Practice Test 6 *(continued)*

11. If you slide figure KLM 2 spaces to the right and 1 space down, what will be the new coordinates of point M?

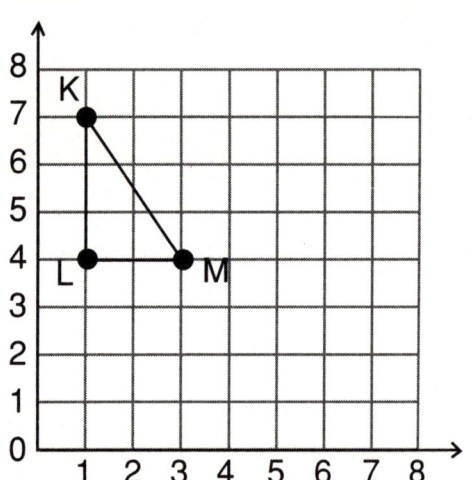

- Ⓐ (3, 4)
- Ⓑ (4, 2)
- Ⓒ (5, 3)
- Ⓓ (3, 3)

12. What is the volume of this tool chest?

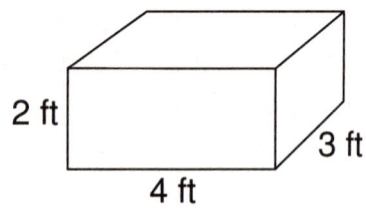

- Ⓕ 9 ft³
- Ⓖ 12 ft³
- Ⓗ 24 ft³
- Ⓙ 48 ft³

For a science project, Carson checked the temperature at 8:00 A.M. each day. He recorded the temperatures on this line graph. Use the graph to answer questions 13–15.

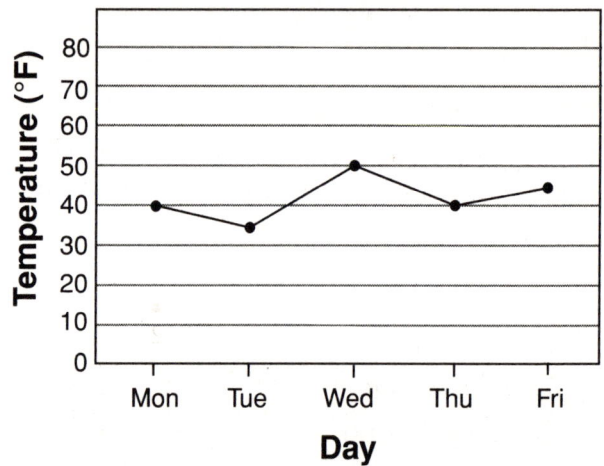

13. What was the temperature on Tuesday?
- Ⓐ 30°F
- Ⓑ 35°F
- Ⓒ 40°F
- Ⓓ 45°F

14. The temperature was the same on Monday and —
- Ⓕ Tuesday
- Ⓗ Thursday
- Ⓖ Wednesday
- Ⓙ Friday

15. What was the difference in temperature from Wednesday to Thursday?
- Ⓐ 10 degrees
- Ⓒ 40 degrees
- Ⓑ 15 degrees
- Ⓓ 50 degrees

GO ON

Practice Test 6 (continued)

16. One side of a square sandbox is 7 feet long. What is the area of the sandbox?

   (F) 98 sq ft

   (G) 49 sq ft

   (H) 28 sq ft

   (J) 14 sq ft

17. Maggie went to sleep at 8:30 P.M. Her clock showed this time when she woke up the next morning. How long did she sleep?

   (A) 9 hr, 15 min

   (B) 9 hr, 30 min

   (C) 10 hr, 15 min

   (D) 10 hr, 30 min

18. Which unit should be used to measure the weight of a package of cheese?

   (F) ounces

   (G) inches

   (H) quarts

   (J) feet

19. Cole spent 2 hours and 15 minutes mowing the lawn. He started at 4:30 P.M. What time did he finish?

   (A) 5:45 P.M.

   (B) 6:30 P.M.

   (C) 6:45 P.M.

   (D) 7:00 P.M.

20. Sherman made 3 quarts of lemonade. How many cups did he make?

   (F) 6 cups

   (G) 9 cups

   (H) 12 cups

   (J) 24 cups

21. Which of these is about 1 meter long?

   (A) an umbrella

   (B) a pick-up truck

   (C) an egg

   (D) a man's shoe

22. There are 20 fifth graders on a school bus. All together, these 20 students probably weigh about —

   (F) 300–500 lb

   (G) 600–800 lb

   (H) 900–1100 lb

   (J) 1400–1600 lb

GO ON

Practice Test 6 *(continued)*

23. A bathtub that is filled with water would hold about how many gallons?

  Ⓐ 3 gal   Ⓒ 300 gal

  Ⓑ 30 gal   Ⓓ 3000 gal

Use a centimeter ruler and the map below to answer questions 24 and 25.

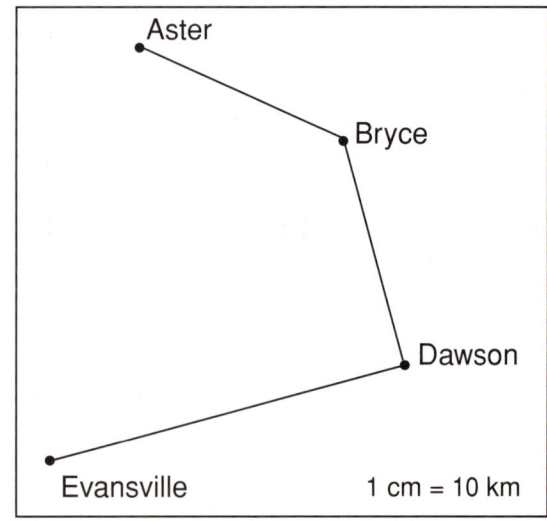

24. What is the actual distance from Aster to Bryce?

  Ⓕ 3 km   Ⓗ 30 km

  Ⓖ 4 km   Ⓙ 40 km

25. Lanny drove from Bryce to Dawson to Evansville. How far did he drive in all?

  Ⓐ 8 km

  Ⓑ 60 km

  Ⓒ 70 km

  Ⓓ 80 km

This graph shows the average attendance at hockey games played by the Cougars each month in 2001. Use the graph to answer questions 26–28.

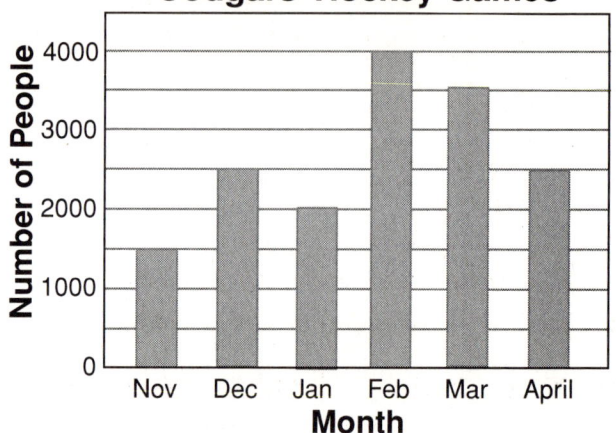

26. In which month was the average attendance highest?

  Ⓕ December   Ⓗ March

  Ⓖ January   Ⓙ February

27. What was the average attendance in April?

  Ⓐ 1500   Ⓒ 2500

  Ⓑ 2000   Ⓓ 3000

28. What was the difference in average attendance between November and December?

  Ⓕ 1000   Ⓗ 2000

  Ⓖ 1500   Ⓙ 2500

STOP

ANSWER SHEET

Practice Test #6

Student Name _____   Grade _____

Teacher Name _____   Date _____

MATHEMATICS

1 Ⓐ Ⓑ Ⓒ Ⓓ Ⓔ	11 Ⓐ Ⓑ Ⓒ Ⓓ Ⓔ	21 Ⓐ Ⓑ Ⓒ Ⓓ Ⓔ	31 Ⓐ Ⓑ Ⓒ Ⓓ Ⓔ
2 Ⓕ Ⓖ Ⓗ Ⓙ Ⓚ	12 Ⓕ Ⓖ Ⓗ Ⓙ Ⓚ	22 Ⓕ Ⓖ Ⓗ Ⓙ Ⓚ	32 Ⓕ Ⓖ Ⓗ Ⓙ Ⓚ
3 Ⓐ Ⓑ Ⓒ Ⓓ Ⓔ	13 Ⓐ Ⓑ Ⓒ Ⓓ Ⓔ	23 Ⓐ Ⓑ Ⓒ Ⓓ Ⓔ	33 Ⓐ Ⓑ Ⓒ Ⓓ Ⓔ
4 Ⓕ Ⓖ Ⓗ Ⓙ Ⓚ	14 Ⓕ Ⓖ Ⓗ Ⓙ Ⓚ	24 Ⓕ Ⓖ Ⓗ Ⓙ Ⓚ	34 Ⓕ Ⓖ Ⓗ Ⓙ Ⓚ
5 Ⓐ Ⓑ Ⓒ Ⓓ Ⓔ	15 Ⓐ Ⓑ Ⓒ Ⓓ Ⓔ	25 Ⓐ Ⓑ Ⓒ Ⓓ Ⓔ	35 Ⓐ Ⓑ Ⓒ Ⓓ Ⓔ
6 Ⓕ Ⓖ Ⓗ Ⓙ Ⓚ	16 Ⓕ Ⓖ Ⓗ Ⓙ Ⓚ	26 Ⓕ Ⓖ Ⓗ Ⓙ Ⓚ	36 Ⓕ Ⓖ Ⓗ Ⓙ Ⓚ
7 Ⓐ Ⓑ Ⓒ Ⓓ Ⓔ	17 Ⓐ Ⓑ Ⓒ Ⓓ Ⓔ	27 Ⓐ Ⓑ Ⓒ Ⓓ Ⓔ	37 Ⓐ Ⓑ Ⓒ Ⓓ Ⓔ
8 Ⓕ Ⓖ Ⓗ Ⓙ Ⓚ	18 Ⓕ Ⓖ Ⓗ Ⓙ Ⓚ	28 Ⓕ Ⓖ Ⓗ Ⓙ Ⓚ	38 Ⓕ Ⓖ Ⓗ Ⓙ Ⓚ
9 Ⓐ Ⓑ Ⓒ Ⓓ Ⓔ	19 Ⓐ Ⓑ Ⓒ Ⓓ Ⓔ	29 Ⓐ Ⓑ Ⓒ Ⓓ Ⓔ	39 Ⓐ Ⓑ Ⓒ Ⓓ Ⓔ
10 Ⓕ Ⓖ Ⓗ Ⓙ Ⓚ	20 Ⓕ Ⓖ Ⓗ Ⓙ Ⓚ	30 Ⓕ Ⓖ Ⓗ Ⓙ Ⓚ	40 Ⓕ Ⓖ Ⓗ Ⓙ Ⓚ

Practice
Test 7

Problem Solving

Practice Test 7

Directions. Choose the best answer to each question. Mark your answer. If the correct answer is *not given*, choose "NG."

1. A restaurant owner puts 7 rolls in a basket for each table. She has baked 135 rolls. What is the greatest number of baskets she can fill?

Ⓐ 17
Ⓑ 18
Ⓒ 19
Ⓓ 20
Ⓔ NG

2. Fred bought 25 pounds of roofing nails.

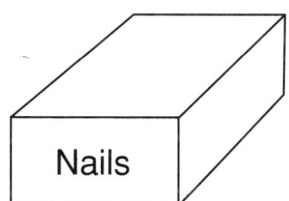

1 pound = 80 nails

How many nails did he buy in all?

Ⓕ 105
Ⓖ 190
Ⓗ 1600
Ⓙ 2000
Ⓚ NG

3. In the town of Acton, the bike path is $5\frac{1}{2}$ miles long. The hiking trail is $3\frac{3}{4}$ miles long. How much longer is the bike path than the hiking trail?

Ⓐ $\frac{3}{4}$ mile

Ⓑ $1\frac{1}{4}$ miles

Ⓒ $1\frac{1}{2}$ miles

Ⓓ $2\frac{3}{4}$ miles

Ⓔ NG

4. Mr. Grande bought 32 pizzas for a school picnic. Each pizza was cut into 8 slices. How many slices of pizza were there in all?

Ⓕ 40
Ⓖ 128
Ⓗ 256
Ⓙ 266
Ⓚ NG

5. Coach Anders ordered 14 hats for the baseball team. The total cost for the hats was $110.60. How much did each hat cost?

Ⓐ $6.80
Ⓑ $7.90
Ⓒ $124.60
Ⓓ $1548.40
Ⓔ NG

GO ON

Practice Test 7 *(continued)*

6. Mrs. Jackson spent $2.79 for orange juice, $13.55 for tuna fish, and $8.90 for bread. How much did she spend in all?
- Ⓕ $22.45
- Ⓖ $23.32
- Ⓗ $24.24
- Ⓙ $25.84
- Ⓚ NG

7. The card shows how many hours Romy worked each day.

Day	Hours
Thursday	4.25
Friday	6.50
Saturday	11.50
Sunday	9.00

How many hours did she work in all?
- Ⓐ 30.25 hr
- Ⓑ 30.75 hr
- Ⓒ 31.25 hr
- Ⓓ 32.35 hr
- Ⓔ NG

8. Mr. White built a brick wall with 8 rows of bricks. Each row was 5 inches high. How high was the wall?
- Ⓕ 3 ft
- Ⓖ 3 ft 4 in.
- Ⓗ 3 ft 8 in.
- Ⓙ 4 ft
- Ⓚ NG

9. Jason started working in the kitchen at 11:30 A.M. He spent $\frac{1}{2}$ hour making salads, 2 hours cooking, and 45 minutes cleaning up. What time did he finish?
- Ⓐ 1:45 P.M.
- Ⓑ 2:15 P.M.
- Ⓒ 2:30 P.M.
- Ⓓ 2:45 P.M.
- Ⓔ NG

10. At the factory, Marco packs 10.4 pounds of shrimp in each crate. <u>About</u> how many pounds of shrimp are in 32 crates?
- Ⓕ 3 lb
- Ⓖ 30 lb
- Ⓗ 300 lb
- Ⓙ 3000 lb

11. In a state park that covers a total of 19,880 acres of land, 4910 acres are open for camping. <u>About</u> how many acres are not open for camping?
- Ⓐ 4000 acres
- Ⓑ 5000 acres
- Ⓒ 14,000 acres
- Ⓓ 15,000 acres

GO ON

Practice Test 7 *(continued)*

12. An average of 3160 cars cross the Gulf Bridge every hour. <u>About</u> how many cars cross the bridge every 12 hours?

- (F) 3600
- (G) 6000
- (H) 24,000
- (J) 36,000

13. Of the 72 students in the school band, $\frac{3}{4}$ are boys. How many band members are boys?

- (A) 18
- (B) 48
- (C) 54
- (D) 60
- (E) NG

14. Workers at a toy factory made 210 dolls in 3 hours. At the same rate, how many dolls will they make in 8 hours?

- (F) 350
- (G) 560
- (H) 630
- (J) 880
- (K) NG

15. According to the scale on a map, $\frac{1}{2}$ inch represents 25 miles. What does 2 inches represent?

- (A) $12\frac{1}{2}$ miles
- (B) 50 miles
- (C) 60 miles
- (D) 75 miles
- (E) NG

The chart below shows the number of colored gumballs Greta has in a bag. Use the chart to answer questions 16 and 17.

Color	Number of Gumballs
Red	10
Yellow	4
Green	6
Purple	5
White	5

16. If Greta reaches in and takes one gumball without looking, what color is she most likely to get?

- (F) red
- (G) yellow
- (H) green
- (J) white
- (K) NG

17. What is the probability that she will pick a purple gumball?

- (A) $\frac{3}{10}$
- (B) $\frac{1}{5}$
- (C) $\frac{1}{4}$
- (D) $\frac{1}{6}$
- (E) NG

GO ON

Practice Test 7 *(continued)*

18. Five girls are standing in line at an ice cream shop. Rita is standing in front of Mary Jo and behind Sue. Lin is standing behind Beth but in front of Sue. Who is first in line?

(F) Sue

(G) Lin

(H) Beth

(J) Rita

(K) NG

19. A total of 200 raffle tickets have been sold, and 50 of those tickets will win prizes. If you have only one raffle ticket, what are the chances that you will win a prize?

(A) 1 in 3

(B) 1 in 4

(C) 1 in 5

(D) 1 in 6

(E) NG

20. Chuck has 6 coins in his pocket. They are all dimes, nickels, and pennies, and he has at least one of each. What is the greatest amount of money he could have?

(F) $0.34

(G) $0.46

(H) $0.52

(J) $0.60

(K) NG

21. On hiking trips, Jane hikes $1\frac{1}{2}$ miles per hour. Which number sentence could be used to find how far Jane hikes in 6 hours?

(A) $6 \times 1\frac{1}{2} = \square$

(B) $6 + 1\frac{1}{2} = \square$

(C) $6 \div 1\frac{1}{2} = \square$

(D) $6 - 1\frac{1}{2} = \square$

(E) NG

22. Charles works 12 hours each week as a baby-sitter. What else do you need to know to figure out how many weeks it will take him to earn $300?

(F) how many children he baby-sits

(G) where he goes to baby-sit

(H) which days of the week he works

(J) how much he earns per hour

(K) NG

23. Risa read 40 pages of a book on Monday, 35 pages on Tuesday, and 60 pages on Wednesday. Which number sentence could be used to find the average number of pages she read each day?

(A) $(40 + 35 + 60) \div 4 = \square$

(B) $(40 + 35 + 60) \times 3 = \square$

(C) $(40 + 35 + 60) - 3 = \square$

(D) $40 + 35 + 60 = \square$

(E) NG

GO ON

Practice Test 7 *(continued)*

24. Ramon ordered these items from a catalog.

> 2 CDs @ $15.00 each
> 3 books @ $4.50 each
> 1 book @ $6.25

What was the total cost of these items (not including tax)?
- Ⓕ $34.75
- Ⓖ $33.25
- Ⓗ $28.50
- Ⓙ $24.75
- Ⓚ NG

25. The table shows how many hours Ike and Tina worked at the church fair on Saturday and Sunday.

	Saturday	Sunday
Ike	$3\frac{1}{2}$	4
Tina	$4\frac{1}{2}$	$2\frac{3}{4}$

How much longer did Ike work?

- Ⓐ $\frac{3}{4}$ hour
- Ⓑ 1 hour
- Ⓒ $1\frac{1}{2}$ hours
- Ⓓ $1\frac{3}{4}$ hours
- Ⓔ NG

26. Mr. Percy made 30 gallons of maple syrup on his farm. He sells the syrup for $12 per quart. If he sells all the syrup he made, how much money will he earn?
- Ⓕ $360
- Ⓖ $620
- Ⓗ $680
- Ⓙ $740
- Ⓚ NG

27. Shelley makes birdhouses to sell. Each birdhouse takes $1\frac{1}{2}$ hours to build and $\frac{1}{2}$ hour to paint. How long will it take her to finish 15 birdhouses?
- Ⓐ 20 hours
- Ⓑ 25 hours
- Ⓒ 30 hours
- Ⓓ 35 hours
- Ⓔ NG

28. At the movie theater, tickets cost $7.00 for adults and $5.50 for children. If 25 adults and 10 children go to a movie, how much money will be spent on tickets?
- Ⓕ $55.00
- Ⓖ $175.00
- Ⓗ $225.00
- Ⓙ $230.00
- Ⓚ NG

STOP

ANSWER SHEET

Practice Test #7

Student Name _____ Grade _____

Teacher Name _____ Date _____

MATHEMATICS			
1 Ⓐ Ⓑ Ⓒ Ⓓ Ⓔ	11 Ⓐ Ⓑ Ⓒ Ⓓ Ⓔ	21 Ⓐ Ⓑ Ⓒ Ⓓ Ⓔ	31 Ⓐ Ⓑ Ⓒ Ⓓ Ⓔ
2 Ⓕ Ⓖ Ⓗ Ⓙ Ⓚ	12 Ⓕ Ⓖ Ⓗ Ⓙ Ⓚ	22 Ⓕ Ⓖ Ⓗ Ⓙ Ⓚ	32 Ⓕ Ⓖ Ⓗ Ⓙ Ⓚ
3 Ⓐ Ⓑ Ⓒ Ⓓ Ⓔ	13 Ⓐ Ⓑ Ⓒ Ⓓ Ⓔ	23 Ⓐ Ⓑ Ⓒ Ⓓ Ⓔ	33 Ⓐ Ⓑ Ⓒ Ⓓ Ⓔ
4 Ⓕ Ⓖ Ⓗ Ⓙ Ⓚ	14 Ⓕ Ⓖ Ⓗ Ⓙ Ⓚ	24 Ⓕ Ⓖ Ⓗ Ⓙ Ⓚ	34 Ⓕ Ⓖ Ⓗ Ⓙ Ⓚ
5 Ⓐ Ⓑ Ⓒ Ⓓ Ⓔ	15 Ⓐ Ⓑ Ⓒ Ⓓ Ⓔ	25 Ⓐ Ⓑ Ⓒ Ⓓ Ⓔ	35 Ⓐ Ⓑ Ⓒ Ⓓ Ⓔ
6 Ⓕ Ⓖ Ⓗ Ⓙ Ⓚ	16 Ⓕ Ⓖ Ⓗ Ⓙ Ⓚ	26 Ⓕ Ⓖ Ⓗ Ⓙ Ⓚ	36 Ⓕ Ⓖ Ⓗ Ⓙ Ⓚ
7 Ⓐ Ⓑ Ⓒ Ⓓ Ⓔ	17 Ⓐ Ⓑ Ⓒ Ⓓ Ⓔ	27 Ⓐ Ⓑ Ⓒ Ⓓ Ⓔ	37 Ⓐ Ⓑ Ⓒ Ⓓ Ⓔ
8 Ⓕ Ⓖ Ⓗ Ⓙ Ⓚ	18 Ⓕ Ⓖ Ⓗ Ⓙ Ⓚ	28 Ⓕ Ⓖ Ⓗ Ⓙ Ⓚ	38 Ⓕ Ⓖ Ⓗ Ⓙ Ⓚ
9 Ⓐ Ⓑ Ⓒ Ⓓ Ⓔ	19 Ⓐ Ⓑ Ⓒ Ⓓ Ⓔ	29 Ⓐ Ⓑ Ⓒ Ⓓ Ⓔ	39 Ⓐ Ⓑ Ⓒ Ⓓ Ⓔ
10 Ⓕ Ⓖ Ⓗ Ⓙ Ⓚ	20 Ⓕ Ⓖ Ⓗ Ⓙ Ⓚ	30 Ⓕ Ⓖ Ⓗ Ⓙ Ⓚ	40 Ⓕ Ⓖ Ⓗ Ⓙ Ⓚ

Practice
Test 8

Computation

Practice Test 8

Directions. Choose the best answer to each question. Mark your answer. If the correct answer is *not given,* choose "NG."

1.
$$\begin{array}{r} 50 \\ \times\ 39 \end{array}$$

Ⓐ 600
Ⓑ 1930
Ⓒ 1950
Ⓓ 2050
Ⓔ NG

2. 7)94

Ⓕ 14 R3
Ⓖ 13 R3
Ⓗ 13 R4
Ⓙ 12 R3
Ⓚ NG

3. 12)580

Ⓐ 46
Ⓑ 46 R4
Ⓒ 48 R2
Ⓓ 48 R4
Ⓔ NG

4. This chart shows the number of plants sold each day at the flower shop.

Plants Sold	
Monday	22
Tuesday	38
Wednesday	47
Thursday	39
Friday	34

What was the average number of plants sold per day?

Ⓕ 180
Ⓖ 45
Ⓗ 36
Ⓙ 30
Ⓚ NG

5. A total of 22 cases of juice boxes were sold at a store.

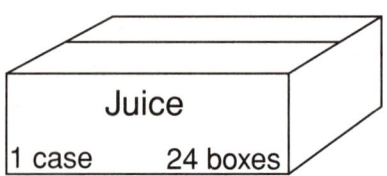

Juice
1 case   24 boxes

How many juice boxes were sold in all?

Ⓐ 96
Ⓑ 428
Ⓒ 524
Ⓓ 628
Ⓔ NG

GO ON

Practice Test 8 (continued)

6. $\frac{5}{6} + \frac{5}{6} =$

    Ⓕ $1\frac{2}{3}$

    Ⓖ $\frac{10}{12}$

    Ⓗ $1\frac{3}{6}$

    Ⓙ $\frac{10}{36}$

    Ⓚ NG

7. $\frac{3}{10} + \frac{1}{5} =$

    Ⓐ $\frac{4}{15}$

    Ⓑ $\frac{2}{5}$

    Ⓒ $\frac{3}{5}$

    Ⓓ $\frac{1}{2}$

    Ⓔ NG

8. $\frac{5}{8} - \frac{3}{8} =$

    Ⓕ $\frac{1}{8}$

    Ⓖ $\frac{1}{2}$

    Ⓗ $\frac{1}{3}$

    Ⓙ 2

    Ⓚ NG

9. $\frac{2}{3} - \frac{1}{2} =$

    Ⓐ $\frac{1}{6}$

    Ⓑ $\frac{1}{4}$

    Ⓒ $\frac{1}{5}$

    Ⓓ $\frac{1}{2}$

    Ⓔ NG

10. A deli offers 5 kinds of sandwich meats, 2 kinds of cheese, and 3 kinds of bread.

Deli Sandwiches		
Meats 5	Cheeses 2	Breads 3

How many different combinations of 1 meat, 1 cheese, and 1 bread can be made?

    Ⓕ 10

    Ⓖ 15

    Ⓗ 24

    Ⓙ 30

    Ⓚ NG

11. At a carnival, these prizes are in a box.

Prize	Number
puppet	8
stuffed animal	10
yo-yo	15
rabbit's foot	7

If you take one prize without looking, what is the probability of getting a stuffed animal?

    Ⓐ $\frac{1}{2}$

    Ⓑ $\frac{1}{4}$

    Ⓒ $\frac{1}{10}$

    Ⓓ $\frac{1}{5}$

    Ⓔ NG

GO ON

Practice Test 8 *(continued)*

12. $\frac{1}{3} \times \frac{3}{4} =$

    (F) $\frac{4}{12}$

    (G) $\frac{3}{7}$

    (H) $\frac{1}{4}$

    (J) $\frac{2}{3}$

    (K) NG

13. $\frac{5}{8} \times \frac{1}{2} =$

    (A) $\frac{3}{8}$

    (B) $\frac{7}{16}$

    (C) $\frac{4}{5}$

    (D) $\frac{1}{2}$

    (E) NG

14. $\begin{array}{r} \$16.75 \\ + \ 4.29 \\ \hline \end{array}$

    (F) $12.46

    (G) $20.04

    (H) $20.94

    (J) $21.04

    (K) NG

15. $3.28 + 4.7 =$

    (A) 7.98

    (B) 7.35

    (C) 7.15

    (D) 3.75

    (E) NG

Use the grid map below to answer questions 16 and 17.

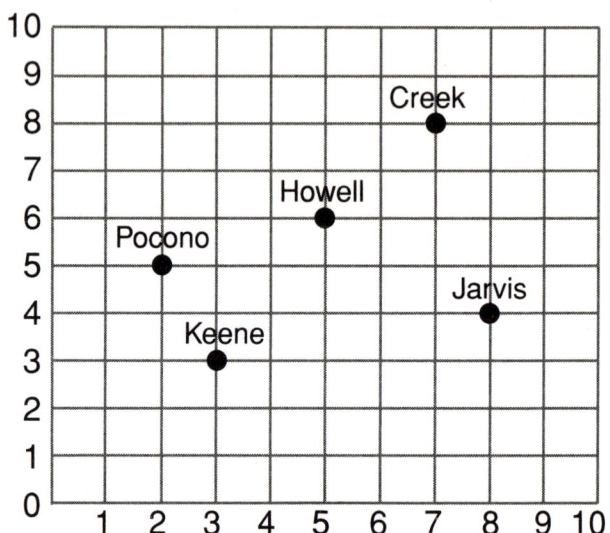

16. What city is located at (2, 5)?

    (F) Keene

    (G) Pocono

    (H) Howell

    (J) Jarvis

    (K) NG

17. Where is Creek located?

    (A) (8, 4)

    (B) (5, 6)

    (C) (6, 8)

    (D) (7, 8)

    (E) NG

GO ON ⇨

Practice Test 8 (continued)

18. $9.1 - 0.7 =$

- Ⓕ 2.1
- Ⓖ 8.03
- Ⓗ 8.4
- Ⓙ 9.8
- Ⓚ NG

19. $\$30 - \$12.75 =$

- Ⓐ $16.25
- Ⓑ $17.45
- Ⓒ $17.75
- Ⓓ $18.25
- Ⓔ NG

20. $5.6 \times 0.4 =$

- Ⓕ 2.24
- Ⓖ 5.1
- Ⓗ 7.4
- Ⓙ 20.24
- Ⓚ NG

21. $18 \times \$3.60 =$

- Ⓐ $54.80
- Ⓑ $60.80
- Ⓒ $64.80
- Ⓓ $65.80
- Ⓔ NG

22. This chart shows the number of new books purchased for the school library each year.

Books Purchased	
1997	40
1998	32
1999	45
2000	18
2001	15

What was the average number of books purchased per year?

- Ⓕ 25
- Ⓖ 30
- Ⓗ 37.5
- Ⓙ 150
- Ⓚ NG

23. Suki wrote this number sentence to solve a problem.

$$(5 \times 4y) \div 4 = 10$$

What is the value of y?

- Ⓐ 2
- Ⓑ 3
- Ⓒ 4
- Ⓓ 5
- Ⓔ NG

GO ON

Practice Test 8 *(continued)*

24. Zoe rolls two number cubes numbered 1–6. What is the probability that she will roll 10 on one try?

- Ⓕ $\frac{3}{35}$
- Ⓖ $\frac{1}{12}$
- Ⓗ $\frac{1}{10}$
- Ⓙ $\frac{4}{9}$
- Ⓚ NG

25. If $(7x + 4x) - 13 = 20$, then what is the value of x?
- Ⓐ 2
- Ⓑ 3
- Ⓒ 5
- Ⓓ 6
- Ⓔ NG

26. Hector wrote this equation to solve a problem.

$$5a - 8 = 37$$

What is the value of a?
- Ⓕ 6
- Ⓖ 7
- Ⓗ 8
- Ⓙ 9
- Ⓚ NG

Use the graph below to answer questions 27 and 28.

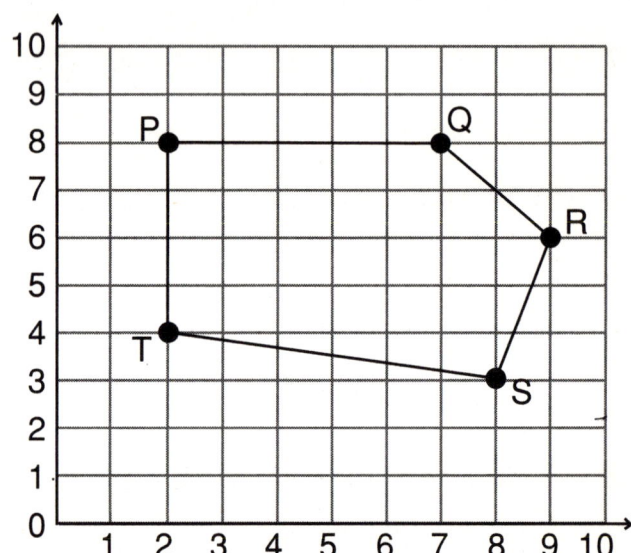

27. What are the coordinates for the location of point R?
- Ⓐ (2, 8)
- Ⓑ (8, 3)
- Ⓒ (8, 6)
- Ⓓ (9, 6)
- Ⓔ NG

28. Which point is located at (7, 8)?
- Ⓕ P
- Ⓖ Q
- Ⓗ R
- Ⓙ S
- Ⓚ NG

STOP

ANSWER SHEET

Student Name _____ Grade _____

Teacher Name _____ Date _____

MATHEMATICS

1 (A) (B) (C) (D) (E)	11 (A) (B) (C) (D) (E)	21 (A) (B) (C) (D) (E)	31 (A) (B) (C) (D) (E)
2 (F) (G) (H) (J) (K)	12 (F) (G) (H) (J) (K)	22 (F) (G) (H) (J) (K)	32 (F) (G) (H) (J) (K)
3 (A) (B) (C) (D) (E)	13 (A) (B) (C) (D) (E)	23 (A) (B) (C) (D) (E)	33 (A) (B) (C) (D) (E)
4 (F) (G) (H) (J) (K)	14 (F) (G) (H) (J) (K)	24 (F) (G) (H) (J) (K)	34 (F) (G) (H) (J) (K)
5 (A) (B) (C) (D) (E)	15 (A) (B) (C) (D) (E)	25 (A) (B) (C) (D) (E)	35 (A) (B) (C) (D) (E)
6 (F) (G) (H) (J) (K)	16 (F) (G) (H) (J) (K)	26 (F) (G) (H) (J) (K)	36 (F) (G) (H) (J) (K)
7 (A) (B) (C) (D) (E)	17 (A) (B) (C) (D) (E)	27 (A) (B) (C) (D) (E)	37 (A) (B) (C) (D) (E)
8 (F) (G) (H) (J) (K)	18 (F) (G) (H) (J) (K)	28 (F) (G) (H) (J) (K)	38 (F) (G) (H) (J) (K)
9 (A) (B) (C) (D) (E)	19 (A) (B) (C) (D) (E)	29 (A) (B) (C) (D) (E)	39 (A) (B) (C) (D) (E)
10 (F) (G) (H) (J) (K)	20 (F) (G) (H) (J) (K)	30 (F) (G) (H) (J) (K)	40 (F) (G) (H) (J) (K)

Practice Test 1 Tested Skills	**Item Numbers**
Numeration and Number Concepts	
Associate numerals and number words	1, 2
Compare and order whole numbers	4, 5
Use place value and rounding	6, 7, 8, 9
Identify patterns (visual, number, odd/even)	3, 10, 11, 12
Estimation	13, 14
Factoring (including GCF) and multiples	15, 16
Identify fractional parts	19, 20
Compare and order fractions (including equivalent)	21, 22
Convert fractions and decimals	23, 24
Compare and order decimals	25, 26
Use a number line (with decimals, fractions, or integers)	17, 18, 27, 28
Apply operational properties	29, 30

Practice Test 2 Tested Skills	**Item Numbers**
Geometry and Measurement	
Identify parts and characteristics of plane and solid figures	1, 7, 8
Recognize symmetry and congruence	3, 6
Identify points, lines, line segments, angles	9, 10, 11
Identify transformations	12, 13
Find perimeter, area, and volume	4, 14, 15, 16, 17
Find elapsed time	18, 19
Use appropriate units of measurement	20, 21, 22, 23
Convert units of measure (standard, metric)	2, 24
Estimate measurements	25
Use scale to determine distance	26
Interpret bar graphs, pictographs, line graphs, tables, charts	5, 27, 28

Practice Test 3 Tested Skills	**Item Numbers**
Problem Solving	
Solve one-step problems using basic operations	1–3
Solve problems involving money, time, measurement	4–9
Solve problems involving estimation and ratio/proportion	10–15
Solve problems involving probability or logic	16–21
Identify steps in solving problems	22–27
Solve multi-step problems	28–30

Practice Test 4 Tested Skills	**Item Numbers**
Computation	
Compute with whole numbers	1, 2, 3, 5
Add and subtract fractions	6, 7, 8, 9
Multiply fractions	12, 13
Add and subtract decimals	14, 15, 18, 19
Multiply decimals	20, 21
Find average, probability, and combinations	4, 10, 11, 22, 24
Solve simple equations	23, 25, 26
Plot points on a coordinate graph	16, 17, 27, 28

Practice Test 5 — Tested Skills | Item Numbers

Numeration and Number Concepts
Skill	Item Numbers
Associate numerals and number words	1, 2
Compare and order whole numbers	4, 5
Use place value and rounding	6, 7, 8, 9
Identify patterns	3, 10, 11, 12
Estimation	13, 14
Factoring and multiples	15, 16
Identify fractional parts	18, 19, 23, 28
Compare and order fractions and decimals	21, 22, 25, 26
Convert fractions and decimals	24, 28
Use number lines (with decimals, fractions, or integers)	17, 20, 27
Apply operational properties	29, 30

Practice Test 6 — Tested Skills | Item Numbers

Geometry and Measurement
Skill	Item Numbers
Identify parts and characteristics of plane and solid figures	1, 3
Recognize symmetry and congruence	4, 6
Identify points, lines, line segments, angles	8, 9, 10
Identify transformations	5, 11
Find perimeter, area, and volume	7, 12, 16
Find elapsed time	17, 19
Use appropriate units of measurement	2, 18, 20, 21
Estimate measurements	22, 23
Use scale to determine distance	24, 25
Interpret graphs, tables, charts	13, 14, 15, 26, 27, 28

Practice Test 7 — Tested Skills | Item Numbers

Problem Solving
Skill	Item Numbers
Solve one-step problems using basic operations	1–4
Solve problems involving money, time, measurement	5–9
Solve problems involving estimation and ratio/proportion	10–15
Solve problems involving probability or logic	16–20
Identify steps in solving problems	21–23
Solve multi-step problems	24–28

Practice Test 8 — Tested Skills | Item Numbers

Computation
Skill	Item Numbers
Compute with whole numbers	1, 2, 3, 5
Add and subtract fractions	6, 7, 8, 9
Multiply fractions	12, 13
Add and subtract decimals	14, 15, 18, 19
Multiply decimals	20, 21
Find average, probability, and combinations	4, 10, 11, 22, 24
Solve simple equations	23, 25, 26
Plot points on a coordinate graph	16, 17, 27, 28

ANSWER KEY

Practice Test 1

Numeration and Number Concepts

1. C	16. G
2. F	17. C
3. B	18. G
4. J	19. C
5. B	20. J
6. F	21. D
7. C	22. G
8. J	23. B
9. C	24. J
10. G	25. A
11. D	26. F
12. H	27. C
13. A	28. H
14. F	29. D
15. B	30. F

Practice Test 2

Geometry and Measurement

1. B	15. B
2. J	16. F
3. A	17. C
4. H	18. F
5. D	19. B
6. G	20. J
7. D	21. A
8. G	22. H
9. C	23. B
10. H	24. F
11. B	25. D
12. J	26. G
13. D	27. C
14. H	28. J

Practice Test 3

Problem Solving

1. B	16. G
2. J	17. B
3. A	18. H
4. G	19. D
5. E	20. J
6. F	21. B
7. D	22. F
8. J	23. A
9. D	24. K
10. F	25. A
11. C	26. J
12. G	27. D
13. B	28. H
14. K	29. E
15. C	30. J

Practice Test 4

Computation

1. D	15. E
2. F	16. F
3. B	17. B
4. H	18. K
5. E	19. B
6. G	20. H
7. C	21. D
8. K	22. H
9. A	23. C
10. J	24. F
11. D	25. B
12. F	26. K
13. C	27. D
14. J	28. F

Practice Test 5

Numeration and Number Concepts

1. B	16. G
2. H	17. D
3. D	18. F
4. G	19. C
5. A	20. J
6. J	21. D
7. B	22. G
8. F	23. B
9. D	24. G
10. H	25. A
11. A	26. H
12. G	27. B
13. D	28. J
14. H	29. C
15. B	30. F

Practice Test 6

Geometry and Measurement

1. D	15. A
2. J	16. G
3. B	17. D
4. F	18. F
5. B	19. C
6. G	20. H
7. C	21. A
8. J	22. J
9. A	23. B
10. F	24. H
11. C	25. D
12. H	26. J
13. B	27. C
14. H	28. F

Practice Test 7

Problem Solving

1. C	15. E
2. J	16. F
3. E	17. D
4. H	18. H
5. B	19. B
6. K	20. G
7. C	21. A
8. G	22. J
9. D	23. E
10. H	24. K
11. D	25. E
12. J	26. K
13. C	27. C
14. G	28. J

Practice Test 8

Computation

1. C	15. A
2. G	16. G
3. D	17. D
4. H	18. H
5. E	19. E
6. F	20. F
7. D	21. C
8. K	22. G
9. A	23. A
10. J	24. G
11. B	25. B
12. H	26. J
13. E	27. D
14. J	28. G